FIRST HIP BOOKS EDITION, JUNE 2026

ISBN: 979-8-234-10129-7

Brought to print by Human Intelligence Publishing

John Nason – Proprietor

Cover art by: Nancy Boudreau

DEDICATION

This book is dedicated to my bro, Rene Boudreau and my twin sis, Nancy Fike. Special Thanks to John Nason for his work on my writing projects. High-fives to Ray Canales, Geoffrey Riggs, Jeff Aldridge, Bart D. Ehrman, Thomas Hertog, Mykel Hawke, and The War On Drugs, "I wanna find what can't be found."

And finally, for all those who value logic and reason above fear and superstition.

ALSO BY DOUGLAS BOUDREAU:

Foraging for Survival: Edible Wild Plants of North America

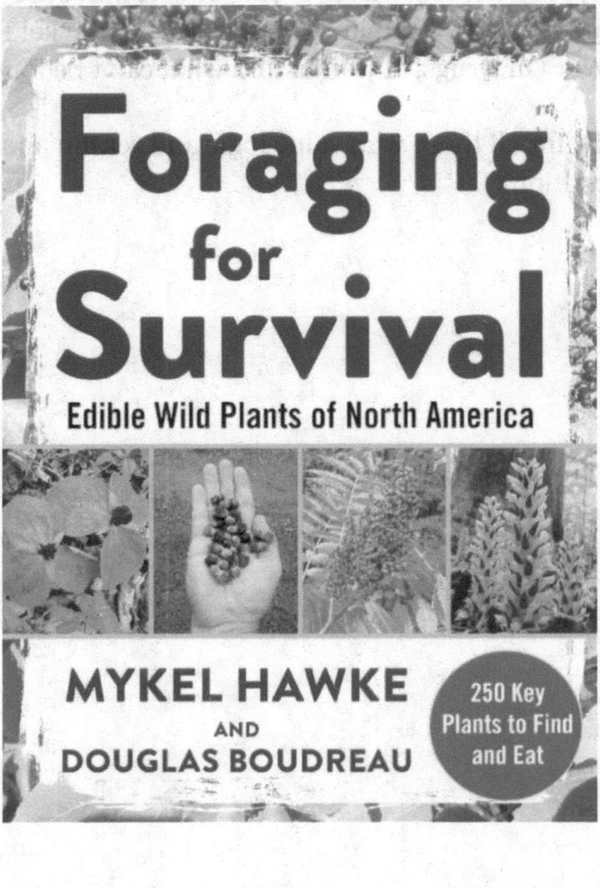

Converging Science and Religion

Evolution is the designed purpose of the universe

Table of Contents:

PART ONE: The Science of IT

PART TWO: The Religion of IT

INTRODUCTION: *Why Do We Exist*

Why do we exist? To amuse some demented deity? I doubt that.

First, the universe had to create the stars that created the elements. That created the means to create the cosmos in the bio-friendly way that it did through the evolving physical laws. As soon as conditions on this planet were established to support life, it emerged in the form of bacteria and molds. This began the cycles of life we are all aware of. Birth, self-animation, procreation, decay and eventually death. The facts tell us that the universe itself evolved to naturally create all the matter and energy necessary to make living things, which appears to be by design.

After many centuries and knowledge acquired for millennia, it was the universe itself and time together that made us. It was not some magic wand waved by some authoritarian magician. Then, the next step to creating us was by somehow developing in us a consciousness of self-awareness; a free will that does not adhere to only the animal instincts that we evolved from. Our subconscious minds can deviate from instinctive behavior with the freedom to choose any course of action for any reason we wish – or no reason at all. The human being is the only life-form here who can make decisions against his own instincts.

These facts tell us that it was the universe which made us and then made us sentient, sapient and self-aware. Where is IT taking us? What appears to be the motivation behind the design of the universe for emerging us from nothing and then adding a unique consciousness of self-awareness? Is it this self-awareness that we got from a fruit from a tree in a "Garden of Eden" that suddenly made us aware of being naked? That might sound silly to modern day thinkers, but that really is where the Holy Bible says as the origin of our unique consciousness in *Genesis* 3:7 and verse 22. Many of you would dispute that because of ancient interpretations of archaic text, but there you have it.

But more to the question at hand would be; how will we further evolve? Increasing paranormal abilities? Space travel? Under present conditions, it will be more than likely our own extinction. However, if we don't completely destroy our species, how will we evolve from here? Are we evolving or devolving now, as the wars and "rumors of wars" continue as always? When will we evolve away from the belligerent, sadistic, and violent confrontations with each other that seems to prevail with every passing century?

If the Designer of the universe had a good reason for this design, it would be for us to eventually let our conscience be our guide and develop a more spiritually benevolent approach to dealing with sharing resources and preventing religions from their destructive behavior.

The core problems on our planet could easily be solved if religion is eliminated as long as the priority of everyone is to help all of humanity to survive and thrive. It is called "Humanism." Join the club. Albert Einstein was a member.

According to the late physics professor Stephen Hawking, the universe itself is as temporary a "living organism" as we are. It was born with a Bang and will eventually run out of energy and time like every living thing within it. He also stated that the universe "appears to be designed" and then in another statement that the universe is "bio friendly".

It's as if the universe was meant to be our home as it evolved for 10 billion years before out planet came into existence. Then, as soon as our planet established an environment suitable for living things, life began. By stating that the universe appears to be designed and that it is bio-friendly, one can assume that it was designed to be bio-friendly. Considering how the stars produce the elements, with the top four elements in the universe being the top four inside of us and a DNA system of evolving life that increases the survival of species by mutating beneficial traits into them that also increases their complexity over time, we can conclude with certainty that the universe was designed to be bio-friendly.

When we see how life evolved in us to acquire a consciousness of self-awareness and advanced communication skills (unlike any other) we can also assume that it was also part of the designed design of everything. A sentient, sapient, and self-aware life-form emerged here from nothing (originally) in this bio-friendly, 4-dimensional, time-space expanding bubble designed by a MIND of some kind and without us knowing why. Scientific consensus is hesitant to accept that evolution is part of a designed design, choosing instead to believe it's just a random chance occurrence that happened here.

Religion can't come up with a logical underlying reason why a MIND of some kind would "create" all this. The claim of us being a god's "children" ignores the fact that the offspring of a life-form is of the same species as the parent.
Existing in a fifth dimensional realm, the MIND behind the design could, theoretically, enter our universe at any moment in history and return to the fifth dimension. That kind of power to manipulate history would be too dangerous in the hands of minds like ours that are still evolving from animals. It is more likely that the Designer has not been involved in human affairs or the natural functions of a universe that need no guiding or directing to evolve. There's more to what's going on here than humanity is aware of at this point. We could say that a "God" of some kind has left us in the "Dark Matter."

"We are still evolving." The late astronomer Carl Sagan famously said that and it has real implications for us to consider. We evolved from animals into evermore complex organisms that acquires a constantly evolving cognition and technology through the acquisition of knowledge with nearly every generation now after a slow start. The question I have been asking is; if there is a MIND from a higher dimensional realm behind this designed design? Has it really been involved in human affairs as religion demands that we "believe", when we are still very much like animals bound and focused on the demands of our physical nature by natural selection, survival of the fittest/richest, which under our system of civilization is far from anything remotely of a "spiritual" nature like the accounts of that iconic first-century Jewish priest and teacher; Jesus of Nazareth.

A real Designer of the Universe has left us in the dark as we evolve into whatever form this design makes us into, provided we don't cause our own extinction trying to survive against the destructive nature of the richest and fittest.
It is logical to assume that a MIND behind the design (if one exists) has not even been communicating with humanity or involved in our choices.

The false prophet and "heretic" called "Apostle Paul" states that our leaders are appointed by "God." At that time, the Jewish people soon became heavily involved in a war for freedom against the Roman empire; which Paul/Saul was a loyal citizen of all along.

In the parable of biblical creation, that primeval tale told in Genesis, we chose the forbidden fruit over the *tree of life*, having the *tree of life* taken away from us which left us on our own to die. (*Genesis* 3:22) The most logical interpretation for the deaths of Adam and Eve, according to a better understanding of story, was for having that tree taken away from them. Any other interpretation is being forced into the story by religion. (*Genesis* 3:23)

According to my alternative reading of that primordial tale, Adam and Eve were never given an immortal soul because the "breath of life" that God breathed into the man's "nostrils" was just oxygenated air to start Adam breathing (evidence of that is elsewhere throughout the Bible). It was in taking the tree of life away from them that doomed humanity to this mortal existence, not having self-existent life within us. That is the proper way to interpret the tale even though the narrative makes confusing statements and we know with certainty that it never happened the way it is narrated.

Furthermore, we can be sure now, and the evidence is there, that Adam and Eve were not the first human being and likely did not exist. This means also that the "original sin" is total fiction. But the human race believes what it wants to believe as the Designer of it all exists in another dimension, waiting to see if we survive the folly of our own ways and evolve true benevolence.

The future is not set in stone even if a consciousness of some kind exists outside of our 3-4 dimensional, time-space expanding bubble of holographic-like, interconnected, entangled, and non-local subatomic particle/waves that have no dimensions of their own at all down to the smallest "Quantum" level. The Designer of the Universe would be an Impersonal Universal Consciousness.

The religious doctrine of the 1st Century Jewish Christology of the Jesus movement and his first followers centered around a doctrine of a future resurrection from the dead like the ones described in Revelation 20. It related what the whole Bible says about life and death when eliminating the pagan-based "Immortal Soul Doctrine" that crept into all Abrahamic religion in those centuries after the first. The references of death throughout is a form of sleep without dreams or awareness of the passage of time.

It is also logical to assume now in consideration of what historical scholars have learned about the early Jewish Christology, that the apocalyptic, messianic Jesus was not born in the manner that Matthew and Luke's Gospels narrate (widely different accounts) because he was a fully human being according to the Jewish Christian groups who were still around.

The Virgin Birth story was made up to make it look like Jesus had been divine (Greek mythology had a lot of that) from his birth. There are many false elements to the Gospels, including how Jesus' family felt about him. James, one of the brother's of Jesus, was a leader among the Jewish population throughout his life and the first (and only for 30 years) leader of the "church" in Jerusalem even though the Gospels and Acts avoided saying it. We know that there is a very anti-Jewish theme in the New Testament and Apostle Paul/Saul was the main reason.

When stripped down to what the Jewish population throughout that first century and later believed about Jesus, a clear outline emerges. When Jesus was Baptized by John the Baptist he believed that the "original sin" was taken away to then be implanted/impregnated with a seed from the tree of life; thus infusing his human spirit with his god's "Spirit" in seed form to grow into the "firstborn," adopted by God, but not actually born until his death. The details for all this are forthcoming. Jesus believed he was the "second Adam." That would completely explain his motivations and the meanings to many of his parables whether he actually performed miracles (embellishments) or came back to life (the evidence is very doubtful).

On June 15, 2018, Late Professor Stephen Hawking had his farewell message beamed into space during a memorial service. (page 266. *"On the Origin of Time: Stephen Hawking's Final Theory"* {2023, Bantam Books}

> "When we see the earth from space we see ourselves as a whole; we see the unity and not the divisions. It is such a simple image, with a compelling message: One planet, one human race. Our only boundaries are the way we see ourselves. We must become global citizens. Let us work together to make that future a place we want to visit."

We are increasing paranormal abilities in the general population. It is as if it was designed that way into our "bio-friendly" (Hawking) universe to emerge and evolve life into us and beyond. If a deity had designed the universe to function naturally for this purpose, it means that the doctrine of refuting the evolution of the species is a huge mistake. If my hypothesis has any truth to it, Stephen Hawking will be visiting that future he hopes for when there will be "time without time."

The Science of **IT** is that life was designed to happen exactly the way it did here by natural processes, not from serendipity; and strictly without any guidance, direction, or interventions by it's "Designer." However, there is a MIND behind the design of the universe.

Renowned Physicist David Bohm on the interconnectedness of subatomic particles, the inner components of atoms:

> "Such quantum wholeness of activity is closer to the organized unity of functioning of the parts of a *living being* than it is to the kind of unity that is obtained by putting together the parts of a machine."

A major key to the science of **IT** is that the atoms of the universe are connected at what appears to be a conscious level down to the smallest of the smallest things in existence, the subatomic particles of atoms, also called the "quanta" of "Quantum physics."

According to one of the first physicists who proposed the *holographic universe theory* (aka "holographic cosmology") and since then repeated by the late Stephen Hawking, the universe functions like a holographic simulation and like a *living being* rather than a mechanical construct or dependent upon some mathematical equation.

However, still yet unrealized, locked into the DNA of all life is a survival driven metamorphic process to enhance the physiology of living things through genetic upgrades whenever species struggle against extinctions and in this way has evolved toward advantages; an evolution in complexity, from simple forms a billion years ago to today's highly complex multi-cellular forms. This evolution of everything (our evolving universe and the emerging and evolving of life wherever the necessary conditions occur) is the key to understanding a logical purpose for the existence of this universe, which originated as a designed design by a MIND of some kind. Not the mind of a god of any religion. For many reasons, as evidenced by the historically complete lack of evidence of a god/God existing and the likelihood that a true MIND behind the design of this universe would be separate from it's *creation.*

According to the apparent design of the universe, wherever and whenever all of the conditions fall perfectly into place in this quantum cosmic construction project — like it did in our little corner of the universe — life can emerge and evolve into sentient, sapient, and potentially self-destructive beings who must drastically increase in complexity away from the greed and violence of their inherited animal instincts to create a stable and peaceful environment for them to transform eventually into benevolence and further the evolution of life to its ultimate designed and designated conclusion. If we don't, this could all be for nothing.

The Religion of **IT** features an alternative interpretation of the purest form of the Adam and Eve "parable" that is amazingly enlightening in its logic of a possibly true overall purpose for the existence of this one and only universe of ours. By "one and only," we suspect now that the *multiverse* theory is not compatible with *general relativity*, therefore indicating that it was a false hypothesis all along. That means that there are no other universes than this one.

However, it will also indicate that, like religion itself, it was designed to distract us from what the amazing truth might actually be, to keep us in the dark while we are still evolving from animals toward a more benevolent nature. Which points toward a very high probability of an overall underlying reason for this one universe existing and producing life from itself.

There is a purpose behind it of which we have never been truly aware of because of science and a few narrative comments by the Genesis writers of the Holy Bible along the prevalent misconceptions of religious beliefs. A fundamental reading of Adam and Eve can perfectly explain that purpose for our existence with the narrative comments by the writers taken out and with proper definitions of biblical words and phrases fully understood from biblical evidence.

Also necessary in understanding the purpose behind this designed design by a MIND of some kind is eliminating the insidious deception of the *immortal soul* doctrine. Once this is understood, the revealing of this underlying purpose of everything results from an accident of chance that the true meaning of life and our existence was there in the parable all along, not from the result of some kind of divine revelation.

Be prepared to jettison most of your indoctrinated theological and Darwinian misconceptions. The most logical theory and explanations possible for the ontological answers of our existence is now within our grasp.

(The War On Drugs):

"We're all just walking through this darkness on our own."

"I wanna find what can't be found."

PART I

The science of IT

1

Stephen Hawking's Final Theory

The sciences have proven to us that conventional religion was a rabbit hole with no answers because the people who made religion did not know where the sun went at night and made up stuff out of a desperate need to believe something; like the beliefs in gods, goddesses, and other unseen entities of mythologies and folk lore. In our world today, science claims that the universe has no direct purpose for existing; that order in the universe emerged from chaotic random occurrences and there is nothing to look forward to beyond this life. The fact that we can question a reason for existing logically means there is an answer.

What are we missing here? There must be something that religion and science are missing because we don't know what we don't know; why do we exist, the ontological question. We have been evolving from animals who evolved from common ancestors and we are still evolving. It isn't magic that we came into existence, no matter what that preacher wants you to believe as he holds out his collection plate.

On the Origin of Time: *Stephen Hawking's Final Theory*, 2023 (Thomas Hertog, pg.245) Thomas Hertog describing the holographic universe theory:

> "In their view, the universe is a kind of quantum information processor, a vision that appears only a hair's breadth from the idea that we live in a simulation… Holography paints a universe that is being continually created. It is as if there is a code, operating on countless entangled qubits, which brings about physical reality, and this is what we perceive as the flow of time. In this sense holography places the true origin of the universe in the distant future, because only the far future would reveal the hologram in its full glory."

> "Top-down cosmology turns the riddle of the universe's apparent design in a sense upside down at the quantum level, the universe engineers its own biofriendliness. Life and the universe are in some way a mutual fit, according to the theory, because, in a deeper sense, they come into existence together."

If our whole universe is void of any reason, purpose, or intention behind its existence because of the claims in recent history of random occurrence or divine guidance being the reasons for its existence, it is not logical. We were created from a universe that cannot ask itself the question "why do I exist" but our conscious minds can.

Stephen Hawking's opinion towards the end of his life had a more metaphysical take on it being of a specific design that functions naturally. In their first meeting some 25 years ago, Stephen asked Hertog (his new collaborator/assistant at the time) "Why is the universe the way it is? Why are we here?"

He had answered those questions the previous moment with a statement that I knew when I read it had more truth to it than the scientific community was comfortable with… "Because the universe we observe appears designed."

What Stephen may have been getting at is my observation as well. To clarify this point, I would ask, why would the universe be "(Hertog) spectacularly well configured to sustain life — if billions of years in the future" if there is no logical reason or purpose of intent behind that function?

(page xiv) "It's almost as if the geneses of life and the cosmos are entwined with each other, that the cosmos knew all along that one day it would be our home."

So you have to ask yourself, as Thomas did...

"What are we to make of this mysterious appearance of intent? It is one of the central questions humans ask about the universe… Stephen felt deeply that cosmological theory had something to say about it."

In my opinion, based on the facts after looking at this deeply for so long, it appears he was right. I shared their hope in the prospect "of being able to crack the riddle of cosmic design."

However, I was disappointed long ago when my research uncovered that "intelligent design" (ID) theory was distracted by the idea that a God was at the controls of the universe, manipulating the design with "His" personal supernatural actions, divine guidance and direction, known as "fine-tuning," also known as the "guiding hand from above" as being the answer to the riddle of cosmic design but without hard evidence for it. My research reveals it was just their desire to use design theory for justifying their theological goldmine of digging up tithes and offerings from an unwittingly unenlightened church-going, world citizens in the business of a lucrative and tax-free "Christian" Industrial Complex.

"Intelligent design" doesn't answer those deepest of questions that Stephen posed to Thomas. In realizing that years ago, I looked for any logic I could find from their theological design theory and came to the conclusion that it is a mistake to equate the cosmic design with a guided and directed (fine-tuned) theory. I also sensed the same detrimental reaction about the Darwinist atheist principle of order deriving from chaotic processes and random occurrences in this "condition that we exist," because we supposedly -just through luck- with no intention of it behind the design to the universe, ended up in a bio-friendly universe for no reason out of the multitude of universes within what was an imagined multiverse, a theory which has lost it's credibility in theoretical physics — in my opinion.

Today we are faced with (Hertog) "critical junctures in science when metaphysical considerations come to the fore, whether we like it or not." Because, at the core of the "quest to grasp what makes the universe just right for life is a humanist question."

That's right! I have concluded from tangible logical evidence that the universe is not here because of a fine-tuned God pulling the strings and pushing the buttons of this "biofriendly" (Hawking/Hertog) living universe within intelligent design's antiquated Creationism. Because the answer to the riddle is a humanist one which excludes the existence of an actively participating god, gods, or any God of any religion. Proponents of intelligent design insist that the universe is being directed, guided, and intervened in by a "strong hand from someplace," (Winston Churchill) otherwise known as a "guiding hand" and usually one described as "from above."

Sorry to bust any indoctrinated fantasies, but there is no guiding hand from above. That has been adequately proven through many fields of scientific discovery, common sense, and our collective global knowledge to this point in our present stage of evolution from animals.

This book is dedicated to answering those questions posed by Stephen Hawking many years ago now, the questions of our own origins; and something I call the why question, also known as the ontological question; that enigmatic reason for being inquired by the subconscious self-awareness in all of us. Why do we exist? Why? For one thing, to evolve in more complexity until we understand it.

Look at it this way, the supreme gift to us by this universe is our ability to question everything and logically examine evidence using reason and critical thought processes.

That should indicate to a critical thinking mind that the answer to the question will require using this ability to question with logic and evidence, which is something that religion discourages.

This bio-friendly universe made us with a conscious self-awareness of intelligent thought. Why did the universe make us this way? Perhaps there is an answer that we have been made to be unaware of, a secret knowledge regarding our underlying, overall purpose for being? Does it have anything to do with the ancient belief of a secret knowledge called "Gnostic"? Or, is it closer to the original belief of Judaism in a resurrection from the dead? What evidence is there that we have some kind of immortal soul and is it logical and therefore credible? Any truth must be logical.

Religion requires us to abandon our cognitive critical thinking skills -that have been developing since we evolved from the animals- to assume their doctrine of beliefs in blind "faith." Will we eventually become truly enlightened to what the overall purpose of our existence is? Yes, we will and science will be part of that journey, as well as parts of the long lost religion of ancient Judaism and original Jesus movement, to reveal what that secret knowledge is at long last. That time is here. We are at the doorstep waiting to be introduced to it. As of this moment, and I have very little doubt about this now.

All life has evolved with changes in complexity (through genetic advancements in its physiology) and there's no reason to think that we have stopped evolving in complexity. The theory of a multiverse and order from chaos must all be jettisoned and replaced with a new perspective and "radically revised understanding of the grand cosmic design" with the wonders of our entangled subatomic particles. Once you take a closer look here at intelligent design theory, you must disregard the fine-tuned and guided aspect of their beliefs.

Why would a God be guiding and directing human affairs while at the same time leaving us completely in the dark about His existence? The most logical and obvious answer is that the MIND behind the design is impersonal and non-existent toward us for a good reason. There is no "HE" behind any religion or religious literature. The sacred documents of religion are completely man-made and I can prove it beyond a reasonable doubt.

One of the most comprehensive examinations of these things are all forth-coming within the pages of this book that required my whole 70 years of asking the why questions to come up with a logical conclusion based on all points of view and facts.

Astrobiology; *Exploring Life in the Universe,* by Lewis Dartnell:

> "Some researchers believe that the emergence of life is effectively inevitable given appropriate conditions; indeed there is evidence that life arose almost instantaneously on our own world. Supporters point to examples where order emerges spontaneously from chaotic systems and patterns of self-assembly and drive toward increasing complexity. The emergence of life may not be due to a lucky non-catalytic reaction of unlikely events but simply be a consequence of the natural development of organic chemistry."

In On the Origin of Time, Thomas Hertog wrote about the Darwinist assumption that life emerging and evolving is a "happy accident." Much of the science community still misunderstands the emergence and evolution of the universe.

He is saying here that it is more likely a specific design. The emergence and evolution of "habitability" in the universe is "a delicate balance of cosmic forces in which any slight change in basic principle properties" [including the chemical elements and the "Elements of the Periodic Table" consisting of subatomic particles] "and forces acting in between them would threaten any habitability in the universe from atoms and molecules to gravity itself."

Thomas Hertog also writes (page 9):

> "So the riddle of design in cosmology is that the fundamental laws of physics are specifically engineered to facilitate the emergence of life. It is as if there is a hidden plot at work that weaves together our existence with the basic rules on which the universe runs. This seems incredible. And it is! What is this plot?"

Did you catch that? That comes from the physicist who worked side by side with Stephen Hawking during the last 20 years of Stephen's life. From the inside flap of the cover-page of his book it reads;

> "Peering into the extreme quantum physics of cosmic holograms and venturing far back in time to our fundamental roots, they were startled to find a deeper level of evolution in which the physical laws themselves transform and simplify until particles, forces, and even time itself fades away. The discovery led them to a revolutionary idea: The laws of physics

> are not set in stone but are born and co-evolve as the universe they govern takes shape."

Physicist Niels Bohr, a colleague of Albert Einstein, on Quantum physics:

> "Everything we call real is made of things that cannot be regarded as real."

He was talking about the subatomic particles within the atoms of all matter and energy. The greatest mystery of science centers on the existence of those particles. It is called "Quantum Physics" because it is the study of the absolutely smallest things that make up our entire existence and we keep finding out strange things about them that don't make much sense to us based on the way we see our reality. If someone claims to understand quantum physics, that person is sadly mistaken or just delusional. However, we are getting closer to figuring it out. Whatever the truth might be, it was designed to keep us guessing to this point as we are still evolving from animals.

And all this, as the inevitable fate of our mismanagement of this planet looms near. The time is now to wake up and ask the right questions. There is a very good reason to consider what it means to be WOKE. The fate of the human race on planet Earth depends on us rising from the physical nature of natural selection to a truly benevolent nature. Being Woke simply means being awakened to the needs of others. To be well informed, thoughtful, compassionate, humble and kind. Eager to make the world a better place for everything that calls it home.

On the Origin of Time, Thomas Hertog:

> "In its most ambitious form, twenty-first-century holographic physics asserts that if only we could decode the hidden hologram, we would understand the deepest nature of physical reality."

What are we? We don't know, we're still evolving from animals. Why does everything exist? All we can be sure of up to now is that the universe and life has evolved from the nothingness that it all started out as. And we are getting closer to figuring it out. We are evolving in complexity toward a higher consciousness, increasing exponentially in the past 50 – 100 years. As our collective conscious cognition increases over time we will find us all interconnected at the subatomic level.

"The Holographic Universe" by Michael Talbot (1991: Harper Collins Publishers; page 299) "From this Ring [Kenneth Ring] concludes that perhaps what we are witnessing is

> "the shamanizing of modern humanity."

But if this is so, why are NDEs [Near Death Experiences] increasing? Ring believes that the answer is as simple as it is profound; what we are witnessing is "an evolutionary thrust toward higher consciousness for all humanity."

Are we getting closer to a point in our evolution in complexity to where we may be able to "decode the hidden hologram" and an evolutionary thrust of increase in the paranormal? New discoveries by the great minds of science, Einstein being a good example, have figured out things that the world reacted to with skepticism — at first. This holographic top-down cosmology in a bio-friendly living universe and evolving complexity in the collective human consciousness are examples of those discoveries of recent years which has drawn skepticism from the world of science, at first.

While it is true that humanity has been evolving in complexity and increasing in our collective consciousness, it has nothing to do with anything that is inherited or in any way immortal/eternal within us. As you will see, the jigsaw puzzle that holds the answer to the mystery of life is designed by its Designer to keep us guessing and blind, stumbling around in the dark, until we improve in nurturing a progressively altered state of mind from the selfishness of natural selection inherent to animals into the humanist benevolent altruism of what an eternally self-existing consciousness would naturally be.

"Altruism" definition, *Webster's New Twentieth Century Dictionary* (1966, World Publishing Company):

"unselfish concern for the welfare of others: opposed to egoism."

2

The Holographic Universe

On the Origin of Time: *Stephen Hawking's Final Theory*, by Thomas Hertog:

> "Theoretical physics discovered while experimenting with hypothetical universes, a "holography" to the universe… Our visible universe as a four-dimensional membrane-like hologram hovering in a fifth dimensional anti-de Sitter space."

Science alert: "Our universe as we know it has four dimensions: the three dimensions of space (up and down, left and right, back and forth), and one dimension of time that keeps us all ticking along."

According to "*wired.co.uk*", A holographic universe means that information that makes up what we perceive as a 3-D reality is stored on a 2-D surface, *including time*."

If the entire universe is "stored on a 2-D surface" it could mean that our complete lives, even our spirit essence of consciousness (if that exists) will be stored there in an inert form once we are dead to return back to the interconnected, non-dimensional, subatomic particles that made us all living organisms when merged together with the elements of matter and energy. If humanity is further evolving in complexity to enable transforming humankind beyond our current stage of evolution with increasing levels of consciousness, it might also mean eventually obtaining access to some degree of control over the functions of this designed holographic universe and of those subatomic particles that behave so mysteriously in experiments that it confounded Einstein.

Scientific American: "Our universe could be a hologram. Much like a 3-D hologram which appears from information encoded on a 2-D surface, our universe's 4-D spacetime could be a holographic projection of a lower-dimensional reality."

"HOLOGRAPHIC; Imagine that everything you see, feel, and hear in these three dimensions emanates from a flat two-dimensional surface." (from Wikipedia and University of South Hampton, Kostas Skenderis: explaining the holographic universe principle)

Now, imagine if this holographic projection affected your/our perception of time and space? Would the approximate age of the projection of our universe, as WE THINK WE know it now, 13.8 billion years, still seem real? If what we see and what we are is a projection, a simulation, that means it is only real to all of us within this time-space expanding bubble that we call the universe.

We are composed of the same elements and molecules that were originally made in the stars of the universe. Does it mean that we are part of that projection? If a hologram is not real in the same sense that we perceive something as real, then what is this universe? That is why physicist Niels Bohr (who lived at the same time as Einstein) said that we are made of things that cannot be regarded as real. Lots of physicists have tackled this holographic cosmology, many more have disregarded it.

David Bohm, a physicist who also lived at the same time as Albert Einstein, had the most amazing take on it.

(Wikipedia) Bohm was a:

> "University of London physicist"
>
> "An American-Brazilian-British scientist who has been described as one of the most significant theoretical physicists and who contributed unorthodox ideas to quantum theory, neuropsychology, and the philosophy of mind. Among his many contributions to physics is his causal and deterministic interpretation of quantum theory known as '*De Broglie–Bohm theory*'.
>
> Bohm advanced the view that quantum physics meant that the old Cartesian model of reality — that there are two kinds of substance, the mental and the physical, that somehow interact — was too limited. To complement it, he developed a mathematical and physical theory of "implicate" and "explicate" order."
>
> "Bohm's main concern was with understanding the nature of reality in general and of consciousness in particular as a coherent whole, which according to Bohm is never static or complete."

On the Origin of Time, Thomas Hertog:

> "Holographic cosmology is a journey deep into the forefront of theoretical physics, interlinking far-flung

> fields, from quantum information to black holes and cosmology, in a tantalizing synthesis that suggests there can be '*time without time*'."

Trying to explain quantum physics and the holographic universe theory in simple terms is an effort of futility for the most part, because of it being so difficult for us to grasp. Furthermore, it stands to reason that it looks like a '*Designer of the Universe*', a MIND behind this designed design, went to great lengths to keep us in the dark until we evolve benevolence. The nature of the universe was also designed to keep us from figuring things out too soon in our evolution from animals to sentience.

The universe itself is not a tangible thing from our point of view, because of not having any dimensions at the subatomic level. As you will see, subatomic particles have no mass to them — no dimensions in their particle-wave/particle form. This means that being part of the universe ourselves confuses our natural intuition of what reality is. What seems real to us does not seem as real when looking at the very smallest, tiniest substance in existence, the quanta, the ingredients of atoms; the subatomic particles.

"The Holographic Universe" (1991, Harper Collins Publishers) pg. 41, Michael Talbot:

> "Classical science had always viewed the state of a system as a whole as merely the result of the interaction of its parts… [instead] the behavior of the parts was actually organized by the whole."

David Bohm, from "*The Holographic Universe*" by Michael Talbot:

> "electrons are not scattered because through the action of the quantum potential, the whole system is undergoing a coordinated movement more like a ballet dance than like a crowd of organized people… such quantum wholeness of activity is closer to the organized unity of functioning of the parts of a living being than it is to the kind of unity that is obtained by putting together parts of a machine."

David Bohm was one of the first physicists to see our universe as a hologram.

From the book, "*The Holographic Universe*" (1991) by Michael Talbot, Chapter intro:

> "One can't help but be astonished at the degree to which [Bohm] has been able to break out of the tight molds of scientific conditioning and stand alone with a completely new and literally vast idea, one which has both internal consistency and logical power to explain widely diverging phenomena of physical experience from an entirely unexpected point of view" "It is a theory which is so intuitively satisfying that many people have felt that if the universe is not the way Bohm describes it, it ought to be." —John P. Briggs and F. David Peat {Looking Glass Universe}.

This could be part of why Hawking/Hertog wrote that the universe operates more like a living organism than a mechanical construct. The more I look at David Bohm's work, the more it appears that it was an integral part of Hawking/Hertog's research that led to their final theory book.

In 1947, Bohm's study at Berkeley on electrons in metals produced the appearance of "seemingly haphazard movements of individual electrons to produce highly organized overall effects." As if it no longer involved just separate particles. The particles knew what the others were doing, like each one knew what maybe trillions of others were doing simultaneously. That is one of the strange characteristics of subatomic particles. Albert Einstein referred to this phenomenon as "Spooky action from a distance." In proven experiments with particles, we know that they are interconnected with all other particles, referred to as being "entangled," and behave simultaneously as if they are all part of the whole, all connected to each other as one, all the way down to the smallest level known to modern humankind.

Another strange phenomenon in particle physics is how our consciousnesses effects the behavior of the particles. The physics riddle of whether a tree falling in the forest makes a sound unless a conscious being is there to hear it, could ring more true than you might suspect. It's as if no particular group of particle waves exist without our attention to them. We know that there is a solar system, a galaxy, and over a trillion galaxies out there because we have observed it with varying degrees of telescopes both crude and the most recently launched super-telescope, the "James Webb Space Telescope."

When we did not know what was beyond the sky, all we could see was lights in the sky at night. We can't say for sure that there are trillions of galaxies now that we know that subatomic particles behave according to who of us (the conscious mind) is watching them at any given moment. I know, it seems strange.

That means to me that our perception of reality makes no sense unless we accept that our reality is a collective of different particles working together as one to provide the functions of what we expect to see. It is an interconnected reality at the subatomic level.

Another strange thing about the insides of atoms is that they have no dimensions. The insides of the elements consist of specific combinations of subatomic particles. More on these things coming. (there is a lot to cover here).

"The Holographic Universe," by Michael Talbot; As we breakdown the parts of matter into its subatomic particles they no longer "possess the traits of objects" (pg. 33) "They have no dimensions of their own, they are like shapeshifters or chameleons. They should be considered both waves ***and*** particles rather than one or the other."

So, is reality truly real or is it just our conscious minds causing a projection of what we see? In a way, it is both; it is real to us because we are all made of the same materials that make up the universe and the Earth. Consider this…

Robert G Jahn (pg. 139, The Holographic Universe):

> "I think we have long since passed the place in high energy physics where we're examining the structure of a passive universe…"
>
> "I think we're into the domain where the interplay of consciousness in the environment is taking place on such a primary scale that we are indeed creating reality by any reasonable definition of the term."
>
> (pg. 54) "The brain is a hologram enfolded in a holographic universe."

David Bohm was uncomfortable with "the other prevailing views in physics… and increasingly troubled by Niels Bohr's interpretation of quantum theory."

He then wrote a text book on the subject and sent copies to Bohr and Einstein asking for their opinions. Bohr didn't respond but Einstein suggested meeting up (as both were at Princeton). For six months they had "spirited conversations"…

> "Einstein enthusiastically told Bohm that he never seen quantum theory presented so clearly. Nonetheless, he admitted he was still every bit as unsatisfied with the theory of Bohm."

What bothered them was it provided no real way of conceiving the basic structure of the universe. If a thing is not real unless we are observing it, it makes the endeavor of science to explain it very difficult. However, David Bohm was so inspired by Einstein to dig deeper that he proposed a new kind of energy field, calling it the "quantum potential" and theorized that like gravity, it pervades all of space.

However, "unlike gravitational fields, magnetic fields, and so on, its influence did not diminish with distance… equally powerful everywhere.." The physics world rejected the idea, "They dismissed it out of hand… virtually all such arguments [against him] were philosophical differences. However, he felt it was important "for scientists to remember that no single cause-and-effect relationship was ever really separate from the universe as a whole."

Could that quantum potential be the same as Einstein's Cosmological Constant that turned out to be Dark Energy and Dark Matter? (this is a legitimate question, that I don't have an answer to yet)

Many scientific breakthroughs in history were initially rejected until more intense research provided evidence for it. Unfortunately, beliefs are addictive. We have religious, philosophical and cultural prejudice, that results in a profound resentment towards any new theory that questions those prejudices. And since western science has devoted several centuries to dismissing the existence of any kind of deist or theist MIND behind the design, they treat those concepts with contempt — not willing to give up their addiction to atheism.

The indoctrinated theist is the same way, unwilling to give up their God delusions.

T.H. Huxley: "Sit down before fact like a little child, and be prepared to give up every preconceived notion, follow humbly wherever and to whatever abyss Nature leads, or you shall learn *nothing*."

The Holographic nature of the universe can explain much of the mysteries of life, the unanswered questions. Like DNA and how our brains can store so many memories in so little space. Physicist John von Newmann calculated that over the course of a human lifetime, the brain stores something like "2.8 x 1020" bits of information.

The evidence of interconnectedness in everything could also reflect the connected nature of a global human collective consciousness at work projecting what we see and hear within the hologram itself. I know it sounds strange. But that is the nature of the true mystery of the ages. It was going to sound strange when you consider that it has been designed to be so deeply mysterious to us in order to keep us in the dark.

The smallest things that we know of are neither solid nor liquid nor gaseous, but can be considered all or neither. This has been well documented. The knowledge of what subatomic particles are is described as "fuzzy" because they have no discernible dimensions. This fact has opened the door to possibilities once considered impossible.

If everything's subatomic particles can change between energy to matter and back to energy, without being destroyed in the process — it stands to reason that a reanimation of subatomic particles of everyone who ever lived would be physically possible by the universal MIND behind its design (not anyone's god/God) of whoever IT wants to "resurrect" when the time comes — likely not until the end of time. Therefore, the possibility exists now within the confines of this hypothesis that religion and science will converge someday with evidence of a true purpose of everything. To evolve our species into an even more complex life-form. The answers to the why questions are here now. We just have to look closely for it. As for who will get to participate in a reanimation (resurrection) of their subatomic particles at the end of time is the trillion dollar question, but we can examine some clues to what the requirements will be:

Empathetic altruism, a truly benevolent nature, a virtual Jesus-like outlook on living is the key.

3

Combinatorial Search Problem

“Genomics Education Programme”: Genomics = DNA.

Put simply, genomics is the study of an organism's genome – its genetic material – and how that information is applied. All living things, from single-celled bacteria, to multi-cellular plants, animals and humans, have a genome – and ours is made up of DNA.”

We have seen claims of so-called Darwinian genetic variations mutating because of natural selection accounting for novel biological form and structure but it implies that it is comparable to genetic functions in DNA blindly stumbling around for no reason.

> Per Wikipedia: “These random mutations (conceived of as copying errors or alterations in hereditary material) supplied the variations upon which natural selection could act and from which new biological form and structure would arise.”

They mean blind copying errors and blind alterations. In other words, stumbling around in the dark, leading to some mutations dying off while others survived and continued to procreate. They are not copying errors, because their genetic function is to enhance the survival of the species being affected. All physical changes over time in the evolution of all species of life happened because of the need of species to change their physiology when extinction threatened to wipe them out. It's the answer to why species increased in complexity since the beginning of life here, from single-celled organisms to us. Therefore, random mutations and copying errors hides the fact that all species of life were engineered to change their physiology when improvements needed to be made for the survival of the species. Look how much humankind has changed in increasingly substantial ways just over the last 5 thousand years in comparison to the last million years. Many evolutionary biologists of Darwinism today agree that copying errors and alterations in hereditary material are insufficient in explaining the beneficial physiological changes and the increasing complexity that took place in the evolution of life.

Evolutionary scientists are realizing this but their preference is to not be vocal about it. After all of the debates between Darwinism and Creationism in recent centuries, these facts were never really addressed because they don't favor either side of the debate and science would rather not give Creationists arguing points that theist-design theorists could use. The design theory I have discovered is a deist based theory, not theist.

Hawking and Hertog concluded that the universe itself is a living entity of biological information and processing systems comparable to actual living things as we know them. Although they did not outright make the statement that it is a designed design by a MIND of some kind, it can be suggested to point in that direction when searching for answers. All scientists are wary of allowing the religious industrial complex to use something they say to jump to a theist conclusion. In other words, to a theist, everything must be the result of a miracle by their god who produced an action to affect an outcome. There is no evidence for that theory that can't be argued against with logic and reason.

DNA works to build life at the molecular level. Billionaire Bill Gates himself once said, "DNA is like a computer program." DNA stores information for building protein molecules that our bodies need to function. Frequent changes occur naturally in this process to enhance the survival of species under the pressure of rising conditions occurring toward the extinction of the species. Like symbols in computer software, DNA molecules carry and convey information depending on how they are arranged in sequences of characters. These carry vital information to all living things on Earth. For every bit of information it carries, there are functional genes, used specifically for certain biological functions in living things. But also, there are "non-functional" genes that have no biological function. The challenge has been in finding the right functional genes for any particular process quickly. It has not been explained thus-far to everyone's satisfaction by the science community just how fast that process works. What is yet to be considered, to my knowledge, is that the interconnectedness of subatomic particles could be the logical reason that biological structures function on a scale faster than lightning because they are all tied together as a whole living organism from the start.

As you will see, there are an infinite number of ways of arranging the characters in DNA of those "nucleotide bases" that carry specific biological information involving every single function of the bodies of every single form of life on Earth as well as whatever life has yet to be discovered. And, as time progressed in evolution, with the number of nucleotide bases multiplied many times over through the course of history, there became a wide gap between the functional genes and the non-functional ones in your DNA system which your body must look through very, very quickly in order to find the particular functional gene that is needed for a specific task — like *lightning* fast. The earliest forms of life a billion years ago had exceedingly less sequences of DNA and as each species mutated, procreated, and changed its physiology, it increased the number of sequences exponentially over a very long period of time.

According to Darwinism's random occurrence theory, finding the right sequence is part of a random process, derived by a chaotic system. You can forget about that nonsense. A random process would not be able to find the right sequences astronomically fast. The old saying in Darwinism goes, a thousand monkeys banging aimlessly on typewriters could eventually write the entire works of Shakespeare. However, how long would that take for it and for the right DNA sequence needed quickly if the process is a completely random one? Of course, not the fraction of a nanosecond that DNA must perform its function in. This is where ID proponents and microbiologists agree that it becomes a needle in a haystack to find any particular DNA nucleotide sequence at any given time. If the process for finding the right one for the job was random, it would take incredibly long in the very brief time available. Every living thing procreates and when they do, they create different sets of genes in DNA sequences based on their parents. When species themselves begin to make physiological changes in the evolutionary process, they must produce new sequences to make those changes. To Intelligent Design (ID) proponents, God is directing, guiding, and manipulating it. They like to throw around the phrase "fine-tuning." It implies an action being taken without saying that it is divine intervention. Which is what they are saying.

As we now know through substantial experimentation for decades, the physical nature of subatomic particles are all interconnected and entangled, as if all the atoms in the universe are connected within a massive and complex holography. Can't we also say the same about the biology of DNA? As we know, particles of all the atoms in existence are functioning as one. The question of DNA is, how does DNA function to find and transfer biological information to where the bodies of living organisms need it in a fraction of a second, among huge amounts of DNA strands of information? It is beyond astronomical! Therefore, the process of the function of DNA is instantaneous and simultaneous, just like the actions of subatomic particles in experiments. Subatomic particles are so completely entangled and interconnected that they react simultaneously with other particles. It's as if everything is all one thing.

Again, the universe is one complete living organism. Hawking/Hertog came to that particular conclusion as well. I contend that it is not random or guided. It was all "*engineered* for *life*" (Hawking/Hertog) by its designed design all along to function as one interconnected and massive living garden within this inflating 3+dimensional time-space bubble of a universe.

"Four Views on Creation, Evolution, and Intelligent Design" by Counterpoints, Bible, and Theology. (This portion written by Stephen C. Meyer, emphasis mine)

> "… the difficulty of a mutational search for a new gene of protein is equivalent to the difficulty of searching for just one combination on a lock with ten digits on each of 77 dials."
>
> "During the entire 3.85-billion-year history of life on Earth, only 1040 individual organisms have lived — meaning that at most only 1040 such opportunities have occurred. Yet 1040 represents only a small fraction of 1077 — only one trillion, trillion, trillionth to be exact. The number 1077 represents the number of possible combinations of DNA nucleotides. It is astronomically huge."

Holography and the interconnectedness of the universe explains why the processes of the functions of the universe like DNA occur instantaneously. It should no longer be a mystery. We should know this now beyond doubt from experiments in subatomic particles. If we could experiment on DNA genetic information storage and retrieval system functions to document just how fast they operate within a living organism, we might get a better idea of just how much our universe and living cells are interconnected as part of the whole. When we break down the molecular structure of a DNA double-helix we find the same basic ingredients in the elements that make our universe's matter and energy: hydrogen, oxygen, carbon, nitrogen, and phosphorus.

Although it seems very much like an advanced form of nanotechnology in many ways, it isn't that. It's interconnectedness at the subatomic level within the functions of the holographic program of the universe.

However, as the ID theory by religion wishes to equate with it, the Almighty Designer himself would have to be sitting at some highly advanced computer keyboard for multiple billions of years adjusting the program and intervening in the physics of the universe to cause nucleotide genes and proteins to choose the right sequences in DNA just to change the physiology of living things. If it isn't magic, and it isn't divine intervention, the only alternative is a design of the universe itself that was engineered, since the beginning, to function the way it does.

To religion, it's magical by a supernatural highly advanced computer tech of many names associated with a "God." It is their God's will, His Omnipotent supernatural force. ID proponents make it look like God is controlling everything. Their ID theory can only be described by any rational person as MAGIC. It was magic, they suppose, that caused this solar system to emerge and configure like it did, emerging life on this planet as soon as conditions allowed its survival. It then adjusted the program to evolve new complex species from old simpler forms until eventually we were created by divine intervention — magic. That is what proponents of intelligent design claim now. Some no longer dispute that we evolved from the same ancestors as primates of what could be called the "Ape kingdom," largely because of the scientific evidence and logic that it happened that way. But to suspend logic for no other reason than to believe the absurdity that we were indoctrinated with by our parents and cultures means we have abandoned logic and critical thinking in exchange for easy fantasies.

In my analysis, this highly advanced holographic program, this "living organism" which is the universe itself, is not being divinely adjusted by its Designer. It's just performing its functions naturally from the designed design engineered by an Impersonal Universal Consciousness, a MIND behind the design. Perhaps it is from a highly advanced kind of blueprint, a template if you will, which was engineered in its initial design to automatically emerge and evolve the universe and life all the way to us wherever all the conditions necessary allows it to happen naturally in this vast universe which recent calculations put it at 2 trillion galaxies and still counting.

It's all much simpler than it might appear to a mind that has been closed to the possibilities. All life as we know it uses the DNA genetic information storage and retrieval system to construct the protein molecules for building and maintaining life's hereditary functions which changes over time, requiring added information for changing each species during the history of life on Earth. Because of being in four dimensions, we are just living beings whose physical bodies evolved from birth to middle age and finally ending upon our demise. The universe as a whole shows the same kind of function in process. Our bodies function as a material biological mechanism which temporarily brings life to our conscious minds, having a unique potential to achieve our level of self-awareness in intelligence and consciousness and dramatically increasing over time.

When humans are willing to rise above the materialism of natural selection and selfishness, our evolution can proceed to the next level. With the entanglement of the physical universe at the quantum level functioning as one living organism, so will everyone who achieves evolving to a much higher level on the evolutionary path of life designed into the universe. We will become telepathic through our conscious minds as we become interconnected with other conscious minds and continue to evolve in ways we can barely imagine. That is the purpose behind the design. A final resurrection is scientifically possible in the future to see who can choose the tree of life over the forbidden fruit, metaphorically. A conscious mind can be a dangerous thing if given the power to control their environments or other conscious minds. A truly intelligent Designer would necessarily require a conscious mind to be enlightened and truly *WOKE*, to be allowed to achieve higher levels of our designed evolution.

ID scientist Stephen C. Meyer with his multiple books promoting the theory of intelligent design:

> "According to Neo-Darwinism, new genetic information arises when random mutations occur in the

> DNA of existing organisms. ‘Random’ here means ‘without respect to functional outcome.” “If mutation is occurring without direction, however, the evolutionary mechanism faces what amounts to a needle-in-the-haystack problem, or what mathematicians call a ‘combinatorial search problem’.”

In mathematics, the term “combinatorial” refers to the number of extremely vast ways a set of objects can be arranged or combined. Notice here that he includes the phrase, “*without direction*” to make indirect reference to their insistence that a God is directing it.

DNA strands grow with changes in lifeforms over time. Each protein nucleotide represents one sequence that can change into what Stephen C. Meyer calls an:

> “Astronomically large number of possible amino acid combinations. approx. 10195. Intuitively, this suggests that the probability of finding even a single functional sequence– i.e. a particular working gene or protein– as the result of random mutational searches may be prohibitively small because of the sheer number of sequences that must be searched by mutations in the available time.”

4

A Biofriendly Universe

If everything is designed to function naturally, there would be no reason for a MIND behind the design to intervene in anything. *Especially* in our affairs, while we are still displaying the mindset of the animals that we evolved from. It is logical.

The universe itself is a "living organism." That's according to Stephen Hawking's statement through his collaborator Thomas Hertog in the recently released book, "*On the Origin of Time: Stephen Hawking's Final Theory*" (2023, Bantum) by Thomas Hertog. Stephen Hawking had changed his mind since his 1998 breakthrough book on theoretical physics, "A Brief History of Time." He had come to believe that he had been looking at it "from the wrong perspective."

My personal theory proposes that the universe is an intelligently designed emergence AND evolution of a highly-advanced but naturally functioning cosmological holographic program. Hawking/Hertog's book talks about "holographic cosmology" as an important step in the future of theoretical physics.

There's a lot to this to reveal that makes sense. There's a lot involved here to straighten out. First, you should know that we evolved and are still evolving. Life evolved by gradually changing in increasing complexity from simple forms of living things to complex structures as species struggled to survive and thrive. Life started as single-celled organisms with minimal DNA and cellular structure, but then changed over time into every living thing on Earth. The process evolved by increasing the information storage and retrieval processes in DNA. The more biological information our DNA can store, the more complex we know life became and can increase with over time. This recipe for beneficial changes was to support any particular life-form's survival as it did for all forms of life that survived to continue as species. Species answered the challenges of survival by changing their physiology (biological structures) without realizing that they were doing it, to whatever enhanced their ability to survive. There is clear evidence in the beaks of birds showing changes that benefited being able to eat certain foods when it was the only kind of food available. They were not aware that it was happening or why. There are slight variations on how and why those changes took place, being all generally the same for many species at the same time and for the same reasons; the threats of extinction. These were natural causes that brought about extinctions for some and a path forward for others. The only thing random about it, was which variations of species would survive to continue their species. However, it was designed to follow a certain path toward complexity.

So how much will we change in increasing complexity from now on? That just depends on if we cause our own extinction event to occur. In other words, as long as our species of humanoid life continues to exist in sufficient numbers to allow our continued survival, we will increase our complexity in ways so remarkable that our present observers of paranormal ability, biology, physics, technology, and evolution will be surprised at how much we can learn and change in the future. It is inevitable, due to the design. Life enhanced its physiology naturally as the need presented itself. Our intelligence and knowledge has been increasing exponentially to this point over the past 100 years compared to where it was.

The reason that DNA in any life-form can process information instantly is because of the entangled, interconnected, and non-local functions of subatomic particles and waves. The universe is all connected at the Quantum level, the smallest possible level that we can detect using our advanced technology.

5

Make Humanity Highly Complex

A "Deist" as defined by Oxford Language, is the belief in "*Deism*":

> "a belief in the existence of a supreme being, specifically of a creator who does not intervene in the universe. The term is used chiefly of an intellectual movement of the 17th and 18th centuries that accepted the existence of a creator based on reason but rejected in a supernatural deity who interacts with humankind."

My theory has a twist to it, a significant difference from the standard definition of Deism. For the sake of argument, if there is ever a component of religion that turns out to be true, in GENESIS 1:26 (NKJV), after human beings evolved but still held the mentality of the animal kingdom without the self-awareness of a consciousness, God said, "Let Us make man in Our image, according to Our likeness." To say "Let us" only implies an intention. Many years ago, I stumbled onto an alternative interpretation of the Adam and Eve tale/parable/folk lore. That story is not a historical event. Ignorant people wanted to believe it that way. They didn't understand the intention behind the story and never have. This alternative theory actually makes the story sound completely logical and even scientifically feasible when using the properly translated and easily interpreted passages from it.

This all becomes the grand design of a jigsaw puzzle that involves both religion and science but was yet to be put together to show a clear picture, until now. The grand design of our existence has left us with a difficult jigsaw puzzle that eludes science and religion to this day. The MIND behind the designed design has left us in the dark, and with good reason to. Our species has not evolved beyond animal instincts.

When we die without being resuscitated, our bodies decay and return to the "dust" of the "earth" (*Genesis* 2:7, 3:19).

However, even smaller than the dust itself are the atoms of the elements of matter and energy which came into existence from inside of stars. Every living or inert thing began as energy waves of subatomic particles before combining with other subatomic particles to become the elements of matter and energy. We were nothing, became matter and energy combined when we were born, and become energy again when we decay back to the soil from which we were taken, the subatomic particles.

All of the materials in our reality are made from atoms that consist of varieties of subatomic particles which, surprisingly and strangely enough, have no dimensions in or of themselves, no mass. This is one of the great mysteries of science. They are not solid, liquid or gaseous; and yet they are all interconnected with each other to some degree throughout the cosmos, as if they are all part of one living entity, because they are.

The science of subatomic particles is indicating that our physical beings and all matter are holographic substances which started as energy created by the stars that were created as they "cooked" (Neil deGrasse Tyson) into all of the chemical elements and the Periodic Table. These matter and energy combinations are made up of interconnected and entangled subatomic particles that are revealing themselves to be self-aware and conscious as a whole in what is called "entanglement."

This is what manifests itself as our 3-dimensional form (the fourth is time itself) when we are alive and conscious. However, the inner being, personality, and character of a consciousness inside of us might be a metaphysical thing, an undetectable form with questionable composition. We are composed of a highly advanced cosmological projection in a program that we can't explain yet. When this spirit (not soul) essence of a human consciousness is alive within a living, breathing body we are sapient, sentient, and self-aware life-forms existing and asking questions in the holographic program of this universe. We don't know what the consciousness of a mind and body working together really is.

We all came into existence when we were born. The mind can do nothing at all without the living breathing body. The living body can do nothing without the mind of the individual, but the consciousness seems to be separate. The religious side of this theory will comparably and amazingly match up with the scientific side. That is why I am calling Part ONE, "The *Science* of IT," and the Part TWO, "The *Religion* of IT."

That essence/consciousness in us seems separate and yet is an integrated part of the program that we are all born with. That consciousness in you IS you. It is not an eternally self-existing organism of any kind. That much will become obvious to you as you read Part Two. Nothing in this holographic simulation of a universe is a self-existing, eternally conscious organism. The so-called "immortal soul" is a huge mistake. Everything emerges and then decays back to its quantum particles. Nothing in our 3-4 dimensional time-space expanding bubble of 2 trillion galaxies is inherently immortal, including us. I call the universe a bubble but once referred to it also as a balloon. However, a balloon expands by having air forced into it. The universe is expanding for some reason that is not reflective of a balloon. A bubble does not expand but calling the universe an expanding bubble is closer to a correct analogy.

Quantum physics, for the laypeople, is the study and close examination of the smallest things in our existence, smaller than the atom itself, which we are yet to fully understand. While the overall science of Physics and specifically astrophysics is the study of the largest things, the cosmos for example; the study of the smallest things are Quantum Physics. We do not understand why subatomic particles know simultaneously what other subatomic particles are doing at the same exact time over long distances away; possibly throughout the entire Universe. However, it provides some answers to the mysteries of life. Answers that neither side of the ontological questions wanted it to be.

We are still trying to figure out the details of the entanglement of atoms, the universal actions of spontaneous awareness by subatomic particles; also calling it "non-local" because it is not restricted to local phenomenon. Everything in the universe down to the smallest things are all entangled by their "interconnectedness." Now that we are trying to understand the interconnected nature of atoms, it has potentially enhanced the possibilities of our technology to levels we once thought were impossible. In my mind, it describes/explains what a highly advanced holographic program might be like that depends on our collective consciousness to exist. It is somehow being influenced by the collective consciousness of our minds. That is all backed up by recent discoveries in particle physics that today's physicists will verify.

Atoms are not solid. Therefore, we ourselves are not as solid as we think we are. Atoms and their particles have no discernible dimensions, so our conscious minds provide the illusion to us of being solid when we observe them. As absurd as it might sound, when you're sitting in a chair, your atoms and the chair's atoms are not actually touching. And not just because of your clothing. The perimeter and composition of atoms and the particles inside of them have no mass, no dimensions — which suggests that they are not real in any practical sense of the word as we understand it. All atoms are composed of subatomic particles within the empty space surrounding the nucleus of the atom. The very perimeter, or what we might call the "surface perimeter" of the atom, is not solid, liquid, or gaseous. It is as if it isn't there. For something as real to us as everything is, to have no dimensions at the smallest level is mind-blowing. We aren't just talking about the inside of the atom being mostly empty space, the surface of the atom itself is not matter, energy or gas. What does it mean? There have been many mind-blowing and confusing discoveries by science over my lifetime. There is evidence that subatomic particles in experiments do not exactly exist until our devices that can detect them are aimed at them. Then they behave in ways that we were not expecting. In other words, the universe doesn't exist unless we are looking at it or consciously aware of it. Without our conscious minds being aware of the universe, the thinking is, that it does not really exist in the concrete way that we perceive it. It has baffled and befuddled even the geniuses of the world of recent history, including Albert Einstein and Stephen Hawking.

As I watch you through the phenomenon of remote-viewing, I see fumes rising from your head. No, just kidding. I can't do remote viewing. But I know people who can. I've even been the target of a remote viewing. It was a very strange experience. I suddenly saw the viewer, someone I know well ("Russell Targ") and had just been with several hours before who has a history of experiments in the paranormal, looking at me from miles away as if he was right there in my room along with his shit-eating grin, sitting in a chair that wasn't really there, for a couple seconds until he suddenly vanished. Yes, I was awake. I had just sat down. It was much more real than any dream.

6

Designed Evolution

Is there a purpose to the existence of all this? Yes, but it has been beyond our grasp to comprehend and we have been misled by the earliest professional liars in history: preachers, along with deceptions from the science of Atheism. We've been stumbling around in the dark by design after evolving from animals into our sentient, sapient, and consciously self-aware minds.

> (per *Wikipedia*) "Eugene Wigner (1902-1995) was a Hungarian American physicist and mathematician who is known for his contributions to quantum physics, chemistry, elementary particles, and the atomic nucleus."
>
> "Science has driven me to a reasonable conclusion in the designed evolution of the universe." ~ Eugene Wigner

One might think -out of the 2 trillion and counting calculated galaxies existing- many are capable of creating and sustaining consciously self-aware, sentient, and sapient physical life as the unique kind that happened here. It is highly probable that other planet Earths exist out there under the highly rare conditions that allows it to happen where it can. It is logical, not magical. If there are other places in the cosmos where human beings have emerged and evolved, not only are they few and far between — but where are they? We would be so far away from any other planet like ours that we cannot know where they are or that they exist. And if it is all by a design, it would be necessary to keep us apart for what must be an important reason — like evolving life from inert matter and energy into benevolent conscious minds. A process that we are still struggling to achieve, but is as obvious a purpose for everything as there could be. To emerge and evolve conscious minds from nothing, into benevolent life-forms over time.

As the universe evolved, it provided the cosmic conditions necessary to be "biofriendly" (Hawking/Hertog) by its natural design with a purpose to produce life. That fact can no longer be denied. There should not be any doubt about it. Stephen Hawking made that absolutely clear in his "*Stephen Hawking's Final Theory*" book (2023) with his collaborator, "Thomas Hertog."

The universe itself is geared toward being a biofriendly habitation for a wide variety of life to emerge and evolve wherever conditions allow it. This process functions naturally within the evolved physical laws of our universe. Even the laws of physics themselves evolved from the Big Bang to shape the cosmos into a breeding ground for life. It is from the natural functions engineered into the designed design of the structure of the universe that is biofriendly. It's not magical, nor is it an anomaly of irregularity and disorder in a hodgepodge of chaos that resulted in serendipity, as believed by the science community for over a century now.

Life was designed to happen exactly the way it did here by natural processes, and strictly without any guidance, direction, or interventions by it's "Designer." Wherever and whenever all of the conditions fall perfectly into place in the configuration like it did in our little corner of the universe, life can emerge and evolve into sentient, sapient, and destructive beings who must rise above their greed and violence to create a stable environment for us to transform into benevolence. If we don't, this could all be for nothing.

Is this a theory of inter-dimensional procreation or just an upgrading humanoid enhancement through this deist designed evolution? This will shine some light on the subject for the first time because of the similarities that finally link modern science and modern religious studies together. A deeper understanding of everything is at our doorstep. Who else will open that door?

7

A New Scientific Revolution

The dawning of a new "scientific revolution" (Hertog) is coming. However, learning about life's interconnected relationship with the physical universe has been impeded by being "fraught with confusion," because "of bafflement at the fact that the supposedly objective, impersonal, and timeless laws of physics are nearly perfectly suited for life."

What that means to me is that some kind of consciousness (not ours) — a MIND behind the design from a higher dimensional plane — is silently behind the scenes, not making any adjustments, guidance, direction, interventions or "fine-tuning" to the design's natural processes that were engineered since the beginning of time. IT's just waiting for us to either advance in complexity and benevolence or failing to evolve any further, as we destroy our own species in mass extinctions as we wait for some kind of miracle from above to save us from ourselves.

Logic suggests that none of the gods, or the perceptions, personifications, and characterizations of any god or "God," would be the Impersonal Universal Consciousness — the MIND behind the design. In order to evolve us from animals we had to develop a conscious self-awareness of our existence. However, we have maintained the animal kingdom's instinct of natural selection, survival of the fittest, through dealing with these physical limitations inherent within us. The purpose of evolving intelligent beings into higher complexity would absolutely require, if not demand, for us to develop a benevolent instinct similar to what is called "*Altruism*." The kind of nature that exemplifies who a Jesus of Nazareth was according to the third-hand narrative and non-historical accounts in the "Gospels" portion of the New Testament of the Holy Bible. You will examine that prospective in Part 2.

Although we are still far from the benevolence stage of development at this point, it would somewhat explain why it was designed for us to be so seemingly left in the dark about the true purpose of evolution; to evolve a higher consciousness within our species. The logic of this is simple. That Designer MIND can't have the greed-filled, apathetic, narcissistic, monsters of history evolving/mutating into a state of being that potentially has access to the higher dimensions. No intelligent designer would allow a universe to exist with nearly or partially omnipotent Hitlers roaming around. However, the concept of a "hell" of ancient mythology is an excuse to believe in the fabricated existence of an immortal soul. There's no such thing as a hell of the Abrahamic religious mythologies.

The English word in the New Testament translated as "Hell" is the result of translating three separate and completely different Greek words from the Latin version of the Greek documents of the New Testament copies preserved over the millennia. People of ancient history did not know where the sun went at night but had vivid and wild imaginations that resulted in their sacred and holy documents. That becomes clear in Part 2, as well as the answers to our deepest unanswered questions about religion. Part 2 will focus on the Judeo-Christian religions but it is a reflection on *all* religions.

8

Time Without Time

Cutting edge science is knocking on the door to fully realizing that the universe as a whole is alive; a "living organism (Hawking)" down to the subatomic level. The subatomic particles in every atom of every element in the universe are all connected to each other the same way that every cell in your body is connected to you. It seems like they are in tune, knowing what others are doing simultaneously over great distances. Einstein was perplexed but curious. They also seem to know when we are using mechanical devices to look at them.

These facts have befuddled the greatest minds in history to this day. Hawking/Hertog suggested that we branch out from the norms of conventional scientific methods of discovery to include any metaphysical aspects that we can extract information from, that will make sense. Imagine what the world would say if the next super-genius declares that scientists need to broaden their search for logical answers to include what many have assumed is magic or supernatural. However, this should not be confused with the pseudoscience of intelligent design. This theory is very different.

The metaphysical can be defined as a MIND from a higher dimension being behind the natural design of the universe, not the fantasy of theists declaring there to be supernatural guidance, direction, and interventions from some mysterious, omnipotent, and totalitarian god/God; an apparently ruthless father figure in the sky.

Linking the physical with the metaphysical has always been in the minds of science because we cannot rule out the possibilities when so many questions remain unanswered. Scientific minds must guard against becoming closed-minded by maintaining an agnostic approach. To be *agnostic* is to simply admit that we don't know what the answer to the ontological question is, so we must keep an open mind.

It also would seem to even the casual observer that everything in our universe operates either like a well-designed machine or a living being. We once believed that it was operating on physical laws that were already established before or at the Creation of the universe. That was not what Hawking and Hertog found in their research in theoretical physics. The physics itself has evolved since emerging; just like us, the galaxy, the universe itself, and life. Today's theoretical physics have been closing doors on old theories of unanswered questions while cracking other doors open to a deeper understanding of entanglement.

Per *Physics of the Universe,* (www.physicsoftheuniverse.com):

> ENTANGLEMENT: {definition} "Nonlocality occurs due to the phenomenon of *entanglement*, whereby particles that interact with each other become permanently correlated, or dependent on each other's states and properties, to the extent that they effectively

> lose their individuality and, in many ways, behave as a single entity."

When I re-read that last quotation I suddenly saw a correlation between what we are evolving into, as I understand it, with the phenomenon of entanglement. As our collective human consciousness evolves toward higher complexity, the goal/purpose could be to, in-effect, become entangled with each other and everything — an interconnection between all of us in our future manifestations, that will lead to the access of higher dimensions and the ability to travel throughout the universe consciously. Maybe so, I don't know. It's certainly possible.

> *Physics of the Universe*: (interconnectedness, aka "entangled") "almost as if the universe at large [as a whole] instantaneously arranges its particles in anticipation of future events. As if acting in unison."

Here is an example of how our universe works. Gravity itself is an example of how the structure of the universe functions. As cosmic debris explodes from dying stars, it floats its way through space and attracts itself toward other debris to merge with and become a part of, becoming interconnected into all forms of the cosmos. The cosmos is teeming with interconnected celestial bodies that formed planets, moons, entire solar systems and galaxies.

I think that's a very good idea for human civilization to work towards, which would give us a better chance of avoiding the mass-extinction event on our horizon. To be interconnected, entangled and non-local throughout the human race with a humanist and altruistic ideology and philosophy; with a political policy and doctrine in the service of humankind instead of serving the fascist, narcissistic billionaires.

If we could accomplish that, the world would become a very peaceful place that -as Stephen Hawking's last words quoted it- we "might want to visit." That is what will eventually *need* to happen if we are to evolve any further in complexity. It all depends on whether we can survive our dumbfounding, idiotic, and haphazard march toward the sixth mass-extinction of planet Earth's long history.

This time, being caused *directly* by us.

9

Fuzzy Quantum Probability Clouds

Scientists of Quantum Physics have discovered credible evidence that Einstein's "Cosmological Constant" was true all along by noticing two invisible and unknown but present forces in empty space that effects gravity and the expansion speed of the universe. It's possible that they are the reason the universe's expansion increases in spite of gravitational forces trying to slow it down. Since we can't see the two opposing and invisible forces we don't know what they are, so we call them "Dark Energy" and "Dark Matter." We have ideas of what their functions are but not what they are, how exactly they function, or why they do what they do. Very much like the universe as a whole, we don't know exactly how it all functions down to its core nuts and bolts.

The Center for Astrophysics: Harvard & Smithsonian; "All the atoms and light in the universe together make up less than five percent of the total contents of the cosmos. The rest is composed of dark matter and dark energy, which are invisible but dominate the structure and evolution of the universe. Dark matter makes up most of the mass of galaxies and galaxy clusters and handles the way galaxies are organized on grand scales. Dark energy, meanwhile, is the name we give the mysterious influence driving the accelerated expansion of the universe. What these substances are and how they work are some of the major challenges facing modern astronomers."

We think that Dark Matter is like the glue that holds the cosmos together and Dark Energy is the reason for the universe still expanding and even speeding up its expansion from what it was just 4 billion years ago. The speed of the universe expanding now is faster than the speed of light. So, even if we could travel the speed of light someday, we won't be able to catch up with the expansion, never reaching the farthest edges of the expanding universe. The evidence points in that direction and is consistent with present astrophysical theory.

If the universe is all interconnected and entangled right down to the subatomic level, it might mean that dark matter and dark energy might also be part of a holographic projection within a highly advanced holographic program.

(The Center for Astrophysics, "When atoms get close enough together, a manifestation of the electromagnetic force, called the Van der Walls force, emerges, that can cause atoms to bond together. This is precisely how molecules form, and the atoms inside molecules are definitely touching, but there is virtually no perimeter of the atom."

> Per *West Texas A&M University*: "Note that the everyday concept of touch (i.e. the hard boundaries of two objects exist at the same location) makes no sense at the atomic level because atoms don't have hard boundaries. Atoms are not solid spheres. They are

> fuzzy quantum probability clouds filled with electrons spread out into waving cloud-like shapes called "*orbitals*" Like a cloud in the sky, an atom can have a shape and a location without having a hard surface."

These facts have some scientists and shamans suggesting that we can change the way reality exists with our conscious minds. Researchers in remote-viewing and other paranormal abilities in humanity are already aware of what we can do with our conscious minds. As we evolve further, we will see more research in that field produce positive results because of the advancing paranormal abilities in the human subconscious. Humanity is experiencing an "evolutionary thrust" in consciousness.

These subatomic things having no dimensions of their own is puzzling. Also puzzling is that inside of the atom is a huge amount of empty space compared to the tiny bit of space that the subatomic particles and the nuclei inhabit. To be exact, the contents of the first atom to exist, the hydrogen atom, is 99.9999999999999% or-so of *empty space*. We should know by now that the universe has been designed to keep us ignorant of the answers to our existential ontological purpose for being.

What is the "top-down" cosmology that Hawking/Hertog wrote about?

We see the timeline of the universe as a progression from the Big Bang to the final fading away of all matter and energy in the vast eons of time ahead. Hawking/Hertog sees a possibility of this process actually moving backward through time from the future conclusion of the universe, where everything actually started from and will reveal itself — along with what it all means. The details are still very fuzzy to me on it. However, if Stephen Hawking and Thomas Hertog saw something to it that made sense to them, it's worth looking very closely at by future scientific study.

> *On the Origin of Time*: "We can now understand at a deeper level how we, a twiglet on the tree of life, together with all other species on our planet, interconnect with the physical universe around us and piece together what it is that breathes life into the cosmos."
>
> Then he quotes Stephen again: "Some people will be very disappointed if in the end there is no ultimate theory," he noted and continued: "I used to belong to that camp. I'm now glad our search for understanding will never come to an end, and that we will always have the challenge of discovery."
>
> 'Stephen appreciated this,' (Thomas Hertog) "If all scientific laws are emergent laws, 'the fundamental' laws of physics included, we are on the brink of discovering a much broader view of nature. These insights tie in with recent developments in a range of scientific disciplines. By jettisoning the idea that we are looking for a unique set of rules, several corners are branching out from studying 'what is' to 'what may be."

The answers will fit together like pieces of a vast and amazing puzzle from Holography, Astrobiology, Genomics,

Microbiology, Astrophysics, Theoretical physics, and other fields of discovery with even some remarkable clues from a nearly 2000-year-old dead religion. As I continued to study these questions of our existence, the pieces of the puzzle fell into place for me.

If the entire universe is a holographic projection within a super-advanced holographic program by an Impersonal Universal Consciousness, a deity of some kind, then we are like holographic characters within a Star Trek starship's "Holodeck" for a purpose that we cannot know at this time. However, I can surmise one thing for sure about the purpose to all this. I imagine it's All You Need Is Love.

> *Carl Sagan*: "We are the product of 4.5 billion years of fortuitous, slow biological evolution. There is no reason to think that the evolutionary process has stopped. Man is a transitional animal. He is not the climax of creation."

> *On the Origin of Time*: "Top-down cosmology turns the riddle of the universe's apparent design in a sense upside down at the quantum level, the universe engineers its own biofriendliness. Life and the universe are in some way a mutual fit, according to the theory, because, in a deeper sense, they come into existence together."

The "ontological dilemma" is our distinctly human ability to ask why we exist, while simultaneously knowing deep down that nobody knows. We do not know the answer to that question. The agnostic and inquisitive approach is the most logical one. Albert Einstein once said, “Never stop questioning.”

This book’s approach is to present this unique and amazing Deist hypothesis on recent discoveries in theoretical physics, for evidence of a physical resurrection from the dead being possible in a future event for a specific purpose of which humankind has been left completely in the dark about thus far, by design, for our own good.

10

Subatomic Particles

Atoms consist of many particles but most notably of three basic “subatomic particles/waves;” “protons, neutrons, and electrons.” There is empty space inside of the atom between the nucleus and an invisible perimeter with those electrons revolving in a floating manner around the nucleus.

> Per *Quora.com*: “Protons and neutrons have orbitals similar to electrons. They are different, partly because of their size, and also the wavelength is much smaller.”
>
> The *American Chemical Society:* “Atoms are made of extremely tiny particles called protons, neutrons, and electrons. Protons and neutrons are in the center of the atom, making up the nucleus. Electrons surround the nucleus. Protons have a positive charge. Electrons have a negative charge. Aug 16, 2023”

All of the elements of the universe are defined by the configuration of these protons, electrons and neutrons within the atom itself. Each element is recognized by the configuration within the atom of these and other subatomic particles with a symbol representing each of the elements labeled by a Latin name {hydrogen, carbon, nitrogen, oxygen}; with the least number of subatomic particles being in the first elements of hydrogen and helium (the gaseous substances) of a cosmic nebula that coalesce to become stars. These were the first elements that formed the stars in which those subatomic particles "cooked" inside of at extremely high temperatures to produce all of the other elements that we know of. The universe itself evolved from those first two elements to the present 118 elements of the Periodic Table, the components of all matter and energy. And we believe that there are more in existence yet to be discovered out there somewhere. It provides us with evidence that the universe evolved it's physics in a growth pattern similar to biological organisms. Life and the universe came into existence together -just not at the same time- and developed over time together as the universe became biofriendly.

> *Chemistry Stack Exchange*, (chemistry.stackexchange.com): "Solids and liquids are bulk systems of interacting atoms/molecules, so no, a single atom cannot be a solid or a liquid. The (hypothetical) situation of an isolated atom in a vacuum would be considered a gas."
>
> *North Coast Photo.com:* "What are pixels in an image? What is a Pixel? A: In digital imaging, a pixel (or picture element) is the smallest item of information in an image. Pixels are arranged in a 2-dimensional grid, represented using squares. Each pixel is a sample of an original image, where more samples typically provide more accurate representations of the original."

The universe appears so huge to us, only because it looks that way from the perspective of our conscious reality in a designed function of the universal hologram itself. Our reality, this space/time continuum, is an extremely advanced holographic projection designed by a MIND of some kind for a purpose that has alluded us until being attainable now in theory, also resulting from its design. The deeper we can look back in time, the deeper into our universe we can see, the larger the universe becomes to us. Every time humankind has achieved the means to look further into the heavens, the larger it all appears. Every single time! It's as if the universe knows when we are watching and shows us what it wants us to see, or what we expect to see. It wants us to keep guessing at what we are and why we exist, like we have been.

Our universe has substance and mass to us, but outside of the Holodeck -in the higher dimensions- we and it don't exist. From the vantage-point of the higher dimensions of the MIND behind the design, our lives are a tiny fleeting moment in the 3+dimensional, time-space bubble where we presently eat, sleep, and exist as individuals for roughly a century (if we are lucky and healthy).

Just like a film at a theater with a 3D projector, what you see from the audience appears to be height, width, and depth even though you're actually seeing a 2-dimensional screen projection with the illusion of 3 dimensions. This might explain why everything looks to us like the universe is an indefinite number of light years in total size with a calculated 2 trillion galaxies in the observable universe, calculated through the discoveries of the James Webb Space Telescope. This means that it's observable to us in our reality by design to keep us guessing in our present stage of evolutionary development. This means that we can acknowledge with modern technology, a calculated potential of there being 2 trillion galaxies in the illusion of a highly advanced holographic projection.

Can we even imagine that amount of galaxies? No, that is where the adjective "astronomical" goes beyond astronomical. Even though the average amount of stars in a galaxy is estimated at one hundred million, there are galaxies known as Super-giants with as many as one hundred trillion stars all orbiting around the center of its galaxy. A habitable zone there is likely, according to our best efforts to understand astrobiology (the study of biological life in the universe).

A number like 2 trillion is beyond our comprehension to fully appreciate. I wonder if a savant in mathematics could comprehend it?

11

Make It So

The answer is "deism" (per *Collins Dictionary*):

> 1. belief in the existence of a God on the evidence of reason and nature only, with rejection of supernatural revelation (distinguished from theism)
>
> 2. belief in a God who created the world but has since remained indifferent to it

IT isn't indifferent. The universe functions naturally by itself with its own processes. But also, we had to go it alone, without the presence or influence from that MIND behind the design as we evolved. IT has to keep us in the dark about why we have to be kept in the dark.

> (Also per *Merriam-Webster's*) "What is the concept of deism? Belief in God based on reason rather than revelation or the teaching of any specific religion is known as deism. The word originated in England in the early 17th century as a rejection of orthodox Christianity."

Also described by (Oxford Language) as Deism, "belief in the existence of a supreme being, specifically of a creator who does not intervene in the universe. The term is used chiefly of an intellectual movement of the 17th and 18th centuries that accepted the existence of a creator based on reason but rejected belief in a supernatural deity who interacts with humankind."

That's it. Finding the overwhelming logic in all this required digging through the sciences of design theory, evolution, theoretical physics, astrobiology, particle physics, microbiology, astronomy, DNA, and general knowledge. We must guard against reading deceptive religious propaganda and being indoctrinated by lies. You can avoid propaganda of religion by only reading and studying the work of critical scholars like Bart D. Ehrman who have no agenda other than finding the truth. Just look for agnostic historians and textual critical scholars while rejecting the nonsensical claims that still prevail in all of religion.

We must reject claims of religious people who hold fast the ancient tradition of using their imaginations and long-established beliefs that doesn't make any sense, from a vast collection of false knowledge. We are not those individuals who for ages didn't know where the sun went after dark. We know so much more now than we ever did and what we know showed us how wrong conventional religion was all these years.

To know all this is truly amazing beyond expression. There aren't enough superlatives to properly address how amazing it is if this hypothesis turns into a very real fact. Something that intellectuals and geniuses have longed for throughout humanity's evolved and semi-civilized history showing up as a rough hypothesis that converges science and a dead religion together being finally revealed here by an introverted loner with no scholarly credentials or background in higher learning.

As the world of science and religion has been stumbling around in the dark looking for a light switch, some of us have now discovered bits and pieces of facts with critical thinking. It can begin to open your eyes as well once you fully grasp contemporary science's Holographic Universe theory and the lost interpretation of Adam and Eve in the garden of Eden parable that will tie it all together.

This universe is a highly advanced holographic program in a theoretical Holodeck, and we are those holographic characters with a metaphysical Captain Picard outside of the Holodeck — impersonal, intelligent, and keeping to the Prime Directive of non-interference in primitive cultures until we finally learn to get along with each other in the true spirit of a civilized civilization.

12

The Greatest Mind of all Time; Albert Einstein

"Learn from yesterday, live for today, hope for tomorrow. The important thing is not to stop questioning." ~ *Albert Einstein*

"We cannot solve our problems with the same thinking we used when we created them."

"Only two things are infinite, the universe and human stupidity, and I'm not sure about the former."

And he was right to not be sure about the universe. The universe is not infinite. It will eventually burn out and end. Albert Einstein was a Deist. That is not atheist or theist. I almost know what he was thinking. Intelligent Design (ID) is a theist doctrine, not deist. ID is the belief that the Designer of the universe is manipulating the physical laws of the universe through guidance, direction and intervention. That is a false premise that has fed into religious delusions, authoritarianism, and bigotry throughout human history; enabling the religious mindset with delusions of grandeur and impeding humankind from evolving into a benevolent nature that IS the true purpose behind the design of the universe.

We are not Christian, Jew, Muslim, or Buddhist, we are all Earthlings — a family of humans.

"Deism" is {Suffolk} – a "Belief in a higher power/supreme being". —{a Belief in an afterlife. [But] rejects revelations as a source of religious evidence. — Oriented toward scientific fact and natural observation, rather than scripture}"

{Wikipedia} "Einstein retained a deistic sense of appreciation of the universe."

{Britannica} "Albert Einstein's religious views have been widely studied and often misunderstood. Albert Einstein said, 'I believe in Spinoza's God.' He did not believe in a personal God who concerns himself with fates and actions of human beings, a view which he described as naive. He clarified however that, 'I am not an atheist' preferring to call himself an agnostic, or a 'religious nonbeliever.' Einstein was closely involved in his lifetime with several humanist groups."'

Oxford Academic: "Einstein showed that time and space are inextricably mixed. Each event requires three space dimensions and a time for its description, so time and space taken together are said to form the space-time continuum. It has four 'dimensions': three for space and one for time."

One of the stunning reveals of "Stephen Hawking's Final Theory," which they hint at throughout, is that humans have a significant role to play: "Our observations have a tangible influence on the universe itself." Our 4-dimensional space/time expanding bubble may not be real at all outside of the bubble.

It has been the intention of whatever MIND created all this to bring humanity to the highest level of evolution possible, to evolve us into a much higher life-form than we are now. In other words, the intention is to "Let Us make man into Our image, after Our likeness."

Something like that seems to me to be the only logical reason there could be. Sound familiar? The parable of Adam and Eve tells a story of the evolution of life and the purpose of life all at the same time. The problem has been how human beings interpreted, translated, and changed it. PART 2 will straighten that out.

13

Dark Matter & Dark energy

Neil deGrasse Tyson, (interview, paraphrased) 85% of the gravity of the universe has an unknown origin. We call it, 'Dark Matter.' Then there's some mysterious pressure acting opposite gravity making the universe accelerate in its expansion over that last 4 billion years. We've known for a long time, there's not enough gravity to stop the expansion, but any amount of gravity ought to be slowing it down, particularly if it was still the result of inflation by the Big Bang. The expansion of the universe is not slowing down, it's accelerating. We don't understand the physics of it to account for that, but we can measure it. So here we are, with two profound aspects of the universe. And if you add up the total presence of these two phenomena – dark matter plus dark energy – it is 96% of everything in the universe. Which means, only 4% of it is what we know about.

We don't know what we don't know because we don't know it yet. That is how we have accumulated knowledge. Not only are we still evolving, we are still learning. We started out as a blank slate like a new-born, void of any tangible knowledge. This includes the human race as a whole and individuals. All we knew was animal instinct in the beginning. Of course it was. That is all we were before we climbed down from the trees to live on the land.

Dark Matter and Dark Energy was first pointed out by Einstein as a “Cosmological Constant” but nobody knew what it was, he wasn’t at all sure about it himself, so he thought he had made a mistake. He announced to the world that it was his biggest mistake and went to his grave thinking it was. It wasn’t a mistake at all. We still don’t know what it is but we do know now that it is there and we call it dark matter and dark energy simply because we can’t see it.

The only kind of scientific approach to any theory, even a "theory of everything,” is the agnostic view and questioning what we think we know. Agnostic is just the word to describe those who do not believe in any God of religion because we don't know that one exists and we haven't seen convincing tangible, empirical evidence of one existing. It doesn’t necessarily mean we are atheist in the sense that we are sure that there is no god/God of any kind. We must maintain questioning in any and all observations about the existence of a MIND behind this grand design.

We think the universe is roughly, at least, 92 billion light years across at present calculations, with at least 2 trillion galaxies and still expanding from the original pea-size primordial atom that it was 13.8 billion years ago just before the Big Bang started the universe inflating. A “light year” describes the speed of how long it takes light to travel through space for a year. Light speed is 186,000 miles per second, equal to 299,792,458 meters per second. We currently have no means of traveling over the speed of light. It is physically impossible. Not even close, for the moment. If we were to even try, any normal human body would die because all matter turns into energy at light speed. Once we understand how to control the functions related to matter and energy transference in holographic cosmology, it would be possible to surpass light speed by a process that turns matter into energy and back to matter in its appropriate form upon arrival to a destination.

It's difficult for us as primordial conscious minds to imagine how fast light-speed is. We can place numbers associated with astronomical calculations, but cannot really wrap our heads around the reality of it. Like the number of galaxies, we cannot envision it in our minds. It is as if it was purposely made that way to keep us ignorant and still growing in cognition. Therefore, the only logical method of science and religion should be to question everything. Science is almost exclusively atheist and religion doesn't want us to question anything that they tell us.

Astrophysicist Neil deGrasse Tyson says, "The universe is under no obligation to make sense to you."

That is quite true, however, I would argue that it actually can make some resemblance of sense to us once we start to put together the jigsaw puzzle. Another name for this theory besides deist designed evolution could be "Astrotheology" or "Astrodeism."

> "*OXFORD RESEARCH*; "Astrotheology",
> "Astrotheology is that branch of theology that provides a critical analysis of the contemporary space sciences combined with an explication of classic doctrines such as creation and a Christology for the purpose of constructing a comprehensive and meaningful

understanding of our human situation within an astonishingly immense cosmos. Within the growing field of Religion and Science, Astrotheology and its sister subject, Astroethics, represent critical responses to the excitement permeating the space sciences, especially astrobiology."

14

Darwinism

To understand what science believes about the functions of evolution, you have to look at Darwinism first. To them, it was a chance to take control of the narrative away from religion by saying that evolution has nothing to do with any kind of designed design theory -whether guided and directed or not- just the random occurrences of order culminating from chaotic processes — an evolution of serendipity.

First, we have to define the science of "*Darwinism*" named after Charles Darwin.

> Per *Google*: "Charles Darwin was a British naturalist who proposed the theory of biological evolution by natural selection. Darwin defined evolution as "descent with modification," the idea that species change over time, give rise to new species, and share a common ancestor."

Darwinian Evolution proposes that simple lifeforms changed over time into all of the lifeforms on Earth by a combination of two simple forms of development.

Natural Selection: lifeforms that worked best in their environments outlived the ones that didn't. Lifeforms that formed better survival skills procreated more than those who could not adapt. Lifeforms that survived global catastrophes and hardships continued to evolve. This is known as, "survival of the fittest." It just depends on the randomness of which species survive extinction events.

Random Mutations: Lifeforms mutated into many forms down to the genetic level in variations of the same forms until they stumbled (the atheist claim) into the forms that worked best for survival at a particular time and environment. According to this, it just randomly changed into many different forms of species according to environmental and hereditary factors. According to this theory, life's evolution is the inevitable result of these random mutations and natural selection. It all happened because of the remote chance that it could, according to the majority of established collective knowledge of scientific inquiry on the subject at this time. It is derived from atheism by atheists, by those who are absolutely positively sure that no god, God, or deity of any kind exists.

Religion has fiercely, violently, and ferociously insisted for thousands of years that a god/God exists and manipulates everything to do "His" will. That's part of the motivation from the science community to choose a random occurrence theory. Both theories, all together, have hid the true nature of the universe being a designed design operating naturally the way it was designed to do all along.

It has been proven beyond a reasonable doubt that life evolved here over the planet's 4-billion-year history from common ancestors mutating physically to adapt and survive changing conditions. Species that could not adapt went extinct. The underlying process required natural selection and somewhat random mutations that is driven by the designed function in living things to survive by any means possible, including, changing their physiology (physical structure) to assist in their survival. Some species who have not needed to adapt to adverse conditions did not change and are still around in the same form they have been for millions of years. The physiological process that enhances species to better survive hardships contributes to their continued survival and procreation.

This process is claimed by atheistic evolution to be a completely random "mutation" occurrence function. Most scientists will not consider the possibility that the physiological increases in complexity for the survival of a species, is a process that was specifically designed into the fabric of living things -like our DNA- in this biofriendly universe. It is a similar process in the evolution of the universe, emerging elements from stars to create the cosmos and pushing advancements in living creatures along as if by a force of nature itself from the beginning. One of the leading advocates for the religious, intelligent design pseudoscience has been a microbiologist...

"The Edge of Evolution: The Search for the Limits of Darwinism" by *Micheal J. Behe*:

> (pg. 233) "is design a philosophical conclusion, a scientific one, or both? Here's where serious opinions diverge. On the one hand, some people think that, if a conclusion implicates an intelligent agent (even merely

> the human mind), or if it threatens to point beyond nature, it's better classified as philosophy. On the other hand, I regard design as a completely scientific conclusion. For many years philosophers have struggled to produce an airtight definition of science, without much luck. But as a rough-and-ready definition, I count as "scientific" any conclusion that relies heavily and exclusively on detailed physical evidence, plus standard logic. No relying on holy books or prophetic dreams. Just the data about nature that is publicly available in journals and books, plus standard modes of reasoning."

It is becoming apparent through the collective knowledge available to us now that there are things about the universe that our species does not understand yet. Unfortunately, his definition of what science is does apply to a theory that produces evidence of a designed design to the universe and evolution but not of a design that is being guided, directed, or fine-tuned. Their evidence does not actually show guidance in evolution. It is better described as naturally occurring processes that were designed into the functions of the universe all along.

15

Intelligent Design is a False Hypothesis

> *C S Lewis:* "Any theory of mind that undermines the validity of human reason cannot be true because you reached the conclusion with reason."

> *Eugene Wigner*, (1902 – 1995), Mathematician: "In science, it is not speed that is the most important. It is the dedication, the commitment, the interest, and the will to know something and to understand it — these are the things that come first."

The purpose of our existence at the moment as defined by this deist designed evolution theory is to get past the internal treachery of our minds willingness to be fooled by vanity and greed, for a greater purpose that humanity has yet to become aware of.

> *Albert Einstein:* "The only incomprehensible thing about the universe is that it is incomprehensible."

> "Intelligent Design" (ID), per *Wikipedia:* "The basic contention of intelligent design is ostensibly scientific: Proponents assert that modern life on Earth could not have developed solely through scientifically established processes of evolution but instead required

> the direct intervention of an "intelligent designer" to produce some or most of the biological phenomena in existence today."

That theory requires the "belief," an assumption, that the design of the universe required direct interventions by its Designer in the process of emerging and evolving this universe, this planet, and us. They insist on this to justify their religious belief in their Bible's authority by a personal God being actively involved with humanity.

> *Britannica*: "Intelligent design, an argument intended to demonstrate that living organisms were created in more or less their present forms by an 'intelligent designer."

They upgraded their Biblical Creationism into calling it "intelligent design" to make it sound more scientific. However, the definition above does not include that fact. The reason for wanting species to originate in their "present forms" is directly because of the first chapter of Genesis called the "Creation Week" and because of how it has been translated and interpreted.

The leading advocates of conventional science and religion have made persistent and vociferous demands in their opposing theories over the last century in literature and debates between "Evolution" and "Intelligent Design." Both sides seem militant and adept at presenting and countering the claims that they present to the world with faulty and fictitious evidence. There is no consensus between the opposing sides. My research has opened the door by combining aspects of both into an all encompassing theory that includes some of their own arguments while exposing the flaws in logic by both sides.

A definition of the word “design” might be a creation of something in the mind, planned and organized for a specific role, purpose, or effect. However, I can’t find anything in books or online for my particular Deist designed evolution theory. However, there are various books, publishers, and online media outlets by atheist and religious organizations that promote their theist intelligent design propaganda, as well as the Darwinian Evolution agenda.

> According to *Wikipedia*: “Intelligent design (ID) is a “Pseudoscience argument for the existence of God, presented by its proponents as "an evidence-based scientific theory about life's origins.” Proponents claim that "certain features of the universe and of living things are best explained by an intelligent cause, not an undirected process such as natural selection. ID is a form of creationism that lacks empirical support and offers no testable or tenable hypotheses and is therefore not science.”

Thank you Wikipedia. But that doesn’t tell the whole story. ID scientists claim evidence that only involves a concept of an imaginary omnipotent supreme entity working the controls of the universe at will; guiding and directing everything. Deist Designed Evolution theory involves proposing a natural universe functioning as it was designed to do from its inception without being guided or directed. It is actually a purely natural process interwoven into the cosmos which has all evolved into what it is now from the Big Bang beginning that it was. As life has emerged and evolved here on this planet, it can happen anywhere in the universe where the exact same conditions occur.

ID proponents often include rhetoric against Darwinism by calling it "undirected" or "unguided," as if a designed design by a MIND must be guided and directed by its Designer — absurdity at it's finest. Those who promote their intelligent design theory, almost exclusively, are the Christian extremists who believe that their "Almighty" God is the intelligent designer and ultimate authority — which makes them the spokes-people for that authority.

These are the same people who want us to ignore the coming mass extinction event that we're causing for selfish reasons, wants to see religions fighting over Palestine to bring about an Armageddon, and wants their religion (themselves) to make the laws that govern everyone, all in anticipating an Armageddon of human imaginations that will only lead to our own extinction. The majority of Christianity will insist that it is their fabled God of the Holy Bible who guides, directs, and intervenes in "His" design to impose "His" totalitarian rule over humankind. That is just a strict authoritarian propaganda tool of controlling the masses through the masked delusion that we must be controlled by fear of the unknown.

The word *Agnostic* from Webster's Expanded Dictionary 1995 Edition quotes "agnostic" as:

> "One who disclaims any knowledge of God or of anything but material phenomena, pertaining to agnostics and their doctrine." That is only one version of being agnostic. Another Webster's Dictionary clearly states the definition of "atheist" is, "One who disbelieves the existence of a God."

Wikipedia has this in reference to 'Agnostic'; "human reason is incapable of providing sufficient rational grounds to justify either the belief that God exists or the belief that God does not exist." Thanks again Wikipedia.

That one seems to me closer to the true meaning but not exactly my specific agnosticism. Because human reason and previously unknown facts actually enables logical evidence capable on the basis of rational grounds to justify the possibility of a Designer of the Universe through a thorough contemporary examination of the science of it and the religion of it. It is apparently an open question which has never truly been examined at the lengths that are being discussed here.

The English biologist Thomas Henry Huxley coined the word agnostic in 1869 and said, "It simply means that a man shall not say he knows or believes that which he has no scientific grounds for professing to know or believe."(Wikipedia)

However, we can now use scientific grounds and logic to acquire a knowledge which has thus far eluded us. The most logical approach to this is how Bart D. Ehrman describes himself, an "Agnostic Atheist" and Albert Einstein referred to himself as a "religious non-believer." I don't know with absolute certainty if there is a god or God of any kind. To accept that we don't know is to be agnostic and that is the beginning of true wisdom. Awareness of one's basic inherited human ignorance is the beginning of wisdom; to admit to oneself that there are things we don't know and can't logically assume to know. To realize this opens the mind to information never considered before. Try it, it works.

A belief in anything is just a matter of having faith that one's belief is the correct one regardless of having reliable evidence; and apparently, even in the face of evidence completely against the belief. When this effects how a few people think it is an anomaly. When it is affecting the majority of the population of the world's top superpowers it can be the most dangerous thing imaginable with global consequences. A potential World War III started by competing religions, using their country's nukes against each other is as dangerous to our planet as anything can be.

Religion has been a major cause of war throughout history, why not now? We can't suppose that they wouldn't go to that level of destruction when it is their hope of a messiah, second-coming, or other such intervention from above that they want to see and their faith tells them it will happen in the event of nuclear war. It's what they want to happen in their lifetimes. I saw someone say that she wanted America to have a dictator who would "make America a Christian nation". I can envision the same lady saying that she welcomes a nuclear war so that her second coming of Christ can happen without further delay. Faith is extremely dangerous.

However, the whole concept of a guiding, directing, and intervening supernatural being in the sky who's watching us and gave us Commandments is obviously man-made to normal common sense and critical thinking. And after all this research has uncovered, it's highly likely that the gods of Judaism, Islam, and Christianity are man-made in order to keep mankind afraid and obedient to a ruling class, along with being profitable in tithes and offerings. Their minds are at ease if they believe that they can buy a stairway to heaven. They can't let anything interfere with that quest, not even the facts behind that "going to heaven or hell" nonsense.

Real science promotes an ideology that first must assume that we just don't know. We continue to look for answers that will make absolute sense or consider the questions unanswerable until such time as they are sufficiently answered.

Brian Cox, (astronomer and genius [probably] with an amazingly logical mind):

> "Without life, the universe is by definition meaningless. It's clear that meaning enters the universe with consciousness and consciousness is a property of living things. And so, without living things there is no meaning. So, I think that let's flip it around. If this is the only planet in the Milky Way galaxy that currently hosts an intelligent civilization, then it's the only island of meaning in a sea of 400 billion stars. And therefore, we have tremendous responsibility, notwithstanding our physical insignificance, to protect this island of meaning. If we mess it up, we will be responsible — if

that's the case — for annihilating meaning perhaps forever in our galaxy."

16

Interconnected Universe

A Designer of the Universe designed us to become empathetic, compassionate, and serve the needs of our fellow Earthlings. Profound knowledge comes with a responsibility to promote the process of evolution of advancing our complexity through empathy. Profound knowledge has profound responsibility.

The Holographic Universe, *Michael Talbot:*

> "Although an electron can sometimes behave as if it were a compact little particle, physics have found that it literally possesses no dimension." "An aspect of quantum reality that Bohm found especially interesting was the strange state of interconnectedness that seemed to exist between apparently unrelated subatomic events."

The Danish physicist *Niels Bohr* pointed out that if subatomic particles only come into existence in the presence of an observer, then "it is also meaningless to speak of a particle's properties and characteristics as existing before they are observed." Makes sense, right?

That really bothered the science community at the time, including Albert Einstein. They take pride in discovering natural phenomenon through the careful observation of tangible evidence. So, if it is the observation itself that is causing the properties of the phenomena, "that create such properties," they cannot explain it and it puts future discoveries in theoretical physics into limbo. After meeting with Albert Einstein and seeing physicist Niels Bohr proposing a solution {that was incorrect}, David Bohm's groundbreaking theoretical hypothesis of Holography and the importance of "interconnectedness" was "swept under the rug" by the science community at that time and many still resist it today. However, anyone can Google or Wikipedia the term "Holographic universe" or simply the word "holography" and read what the current research and opinions are. Study up on where the science has gone with it. It would be crucial to understand these things moving forward in our evolution.

The Holographic Universe (*Talbot*):

> "During this same period of his life, Bohm also continued to refine his alternative approach to quantum physics. As he looked more carefully into the meaning of the quantum potential, he discovered it had a number of features that implied an even more radical departure from orthodox thinking. One was the importance of wholeness. Classical science had always viewed the state of a system as a whole as merely the result of the interaction of its parts."

So, if microbiology is displaying an interconnectedness and an "irreducible complexity" — which means that living cells cannot change their molecular structure randomly because they must change in ways that continue the functioning of the cell without interruption — then everything within the biological form is interconnected in such a way that functions together as a whole, not as separate subatomic particles, microbes, molecules or the system of DNA and genetics.

Evolving a consciously aware and interconnected universe is the reason why evolution has changed the physiology of all life toward beneficial complexity over the course of geological history with its design to function naturally. Not a random occurrence or a divinely intervened guidance and direction.

This "quantum potential" (*The Holographic Universe*) of the whole is looking like individual particles of atoms acting in unison, like watching an aquarium on two separate monitors of two separate video cameras showing the same aquarium at different angles. As the fish move around, the monitors look like they are different fish in a different aquarium, but you will notice that their movements are in sync. They are the same fish, just viewed from different angles. This helps explain why atoms and subatomic particles (electrons, protons, quarks, etc.) seem to work together as if they all know what actions are required simultaneously to function everything in existence.

Logical evidence of a designed design is knocking on our door but both sides, atheism and theism, refuse to acknowledge it. No surprise there.

> Quantum Physics Lady: "Nonlocal Interconnectedness; quantum particles can "know" the states of other quantum particles, even at great distances, and

correlate their behaviors with each other instantaneously..."

The Physics of the Universe: "Nonlocality and Entanglement — Nonlocality occurs due to the phenomenon of entanglement, whereby particles that interact with each other become permanently correlated, or dependent on each other's state, even when separated by large distances (potentially even billions of light years) almost as if the universe at large [as a whole] instantaneously arranges its particles in anticipation of future events."

"Nonlocality occurs due to the phenomenon of entanglement, whereby particles that interact with each other become permanently correlated, or dependent on each other's states and properties, to the extent that they effectively lose their individuality and, in many ways, behave as a single entity. The two concepts of nonlocality entanglement go very much hand in hand."

17

The Double-Slit experiment

A holographic universe would explain how the universe looks so infinitely huge to us and once appeared to many scientists to be a multiverse. Scientists of Astrophysics like Neil deGrasse Tyson are revealing how subatomic particles behave in a particle accelerator collider. When a detection device is observing how these subatomic particles move through two separate openings, (the double-slit experiment) the pattern is different when it is not being observed.

The double-slit experiment reveals a phenomenon involving shooting photons of light, subatomic particles of the atom, through a double slit devise in a Particle Accelerator, but not for the purpose of colliding the particles. The two slits are side by side. The photons travel one by one, some through one slit and some through the other. We expect to see a constant beam of light through the two slits with a pattern of light on the wall behind the devise showing the light beams going through the two slits. The pattern of light changes depending on whether it is being observed or not.

Einstein said, "I, at any rate, am convinced that "God" does not throw dice."(Wikipedia)

However, he was talking about the theist version of a guiding, directing, and intervening god throwing dice. He called himself a deist and humanist because he suspected that a MIND of some kind had set the wheels of the universe in motion without interference. Keep in mind that Albert Einstein was very much a deist humanist like myself. The statement above was pointing out his disagreement with Darwinism concerning random mutations being analogous to rolling dice.

Consciousness is what brings our reality into existence. This is where science and mysticism converge. The following paragraph is from

> "*Essay — The Illusion of Reality*" Issue 37, by Christopher Bek (2018)
>
> "Michael Talbot wrote in his 1993 book, "Mysticism and the New Physics" to argue that we are on the precipice of a monumental paradigm shift regarding the ultimate nature of reality. To paraphrase Talbot "Nowhere is the term paradigm shift (i.e. A radical change of perspective that is sweeping the Western world) advancing so rapidly as in physics. Mystics, Tantrics and Idealists have always propounded the idea that reality is an illusion. Now quantum theory is putting forward arguments that reinforce this belief."

Unfortunately, for most people, their judgment of others without empathy, compassion, or altruistic benevolence will prevent them from obtaining the necessary enlightened awareness (being WOKE) and embrace the selfless, humbled, nature that will be required to participate in the next stage of our evolution. "For all is vanity and grasping for the wind."

18

Unintelligent Design

A popular scientist of the Intelligent Design community, considered an authority and prestigious author on the subject, is *Stephen C. Meyer* — who stated from his book, "Creation, Evolution, and Intelligent Design" {page 207}:

> "Though intelligent design is not based upon religious belief, it does affirm a key tenet of a biblical worldview — namely, that life and the universe are the products of a designed intelligence — an intelligence that I, and other Christians, would attribute to the God of the Bible. Nevertheless, the theistic implications of intelligent design are not grounds for dismissing it. Scientific theories must be judged by their ability to explain evidence, not by whether they have undesirable implications."

Those who espouse intelligent design all have similar opinions without concrete "evidence" to explain it without twisting the facts around and tying their explanations in knots with biblical Creationism. They insist on the universe being guided, directed, and manipulated by the supernatural power of their god, as people have been doing forever.

Moreover, their answer to the riddle of design is one of magical interventions — guidance and direction — by the Designer through "His" continuous supernatural monitoring of the universe and taking action to make adjustments to the holographic program to accomplish His authoritarian/totalitarian purpose of stroking His own ego.

It's usually the Jewish, Islamic, or Christian God who they claim is that guiding and directing MIND behind the design of the Universe.

That same "God" of many names who said in one of the "Ten Commandments", "don't take my name in vain." The same God who went by many names, such as; "Adonai, El. Elohim, Eloah, Mighty One, Mighty Ones, The Almighty, LORD, Lord, God, LORD God, YHWH, Jehovah, Yahweh, Allah, the "I AM" and other names for supposedly the same personage.

> *Thomas Edison* is quoted (in the Times Herald) saying, "I do not believe in the God of the theologians, but that there is a Supreme Intelligence, I do not doubt."

Darwinists claim the only known theory that could solve the mystery of our existence is Darwinism. At the same time, they are astounded at how much the universe and living organisms are "unmatched by the elegant efficiency of their apparent design." (*Richard Dawkins*, The Blind Watchmaker) To reflect on the "elegant efficiency" of a "design" is analogous of the ultimate clock or computer having an elegant efficiency of an apparent design. This universe is all intentionally designed by a MIND of some kind like the clock and the computer.

Stephen C. Meyer PHD (in Lee Strobel's "The Case For A Creator): "... if it's true there's a beginning to the universe, as modern cosmologists now agree, then this implies a cause that transcends the universe. If the laws of physics are fine-tuned to permit life, as contemporary physicists are discovering, then perhaps there's a designer who fine-tuned them. If there's information in the cell, as molecular biology shows, then this suggests intelligent design. To get life going in the first place would have required biological information; the implications point beyond the material realm to a prior intelligent cause."

The prior intelligent cause is how the Big Bang was engineered and orchestrated from before the beginning to function independently of a MIND of any kind. It is a deceptive practice in religion to put a theist spin on everything, regardless that evidence for their claim is available for examination.

The inner functions of biology down to the smallest molecules and DNA strands is a Deist designed process functioning naturally at the subatomic level. It is not the actions of a fine-tuning God in real time changing the cell's structure by some magical guiding hand that doesn't exist. What they are proposing is still Creationism but with lots of deceptive spin to it.

19

The Blind Watchmaker

> *Richard Dawkins*: "I think a case can be made that Faith is one of the world's greatest evils, comparable to the smallpox virus but harder to irradiate... I'm talking about faith, where "faith" means belief in something without evidence. Because, if you believe in something without evidence, then you can justify anything in your own mind. You're no longer vulnerable to somebody coming back at you saying, 'hang on a minute, let me argue the case.' If you believe it without evidence, which is what faith is, then you don't argue the case, you say, 'No, I'm not arguing the case, this is my faith, it's mine. It's private, I don't descent from it, I don't retreat from it. You're just going to have to accept it now.' Now that is evil."

That is from one of the leading atheists of the world. In his book, "*The Blind Watchmaker*" Dawkins admits some doubts about the conventional belief in the claim of cumulative selection, having been recently discredited over what the accumulation of fossil records are showing us.

"Cumulative Selection" is the claim by standard Evolutionists ever since *Charles Darwin* (1809-1882) introduced his findings of changes occurring over very long-drawn-out periods of Geological time — strictly by natural selection and random mutations of genetic copy errors — according to standard evolutionary science. It was proposed that it was a slow cumulative system of evolving life from the simple early forms to the complex multicellular forms that it eventually became by a randomly occurring process. A randomly occurring process would show nothing more than very long periods of slow changes in species with little to no variation.

Dawkins knows we have since found out from the fossil record that evolution wasn't exactly a slowly accumulating process when we see the opposite; long gaps of stagnation with relatively short bursts of very dramatic changes. That is why intelligent design's "guiding, directing, and fine-tuning" theory is so appealing to religious apologists. The science community has always been over-confident in their own theories, which left them vulnerable to being wrong. They all want to be right whether it makes sense or not and whether the evidence supports it or not.

By "random" in the term "random mutations," it's implying that life's inner functions all stumble around blindly by chaotic processes to change DNA randomly and the changes that survived continues on; as opposed to the random changes that doesn't help a species survive and becomes extinct. Seriously, that is what is implied by it. To evolutionary science, it is serendipity based on which species are lucky enough to change randomly toward ways that adjust their DNA to deal with the mass-extinction events that threatens their existence. They're tying themselves in knots trying to make sense of it, but it doesn't work.

> *Springer Link:* "Cumulative selection is a powerful process in which small changes accumulate over time because of their selective advantage. It is central to a gradualist approach to evolution, the validity of which has been called into question by proponents of alternative approaches to evolution."

The fact that life emerges anywhere that it can and advances in complexity, modifying its own physiology with DNA mutations to make survival easier, is no random occurrence. It was designed into the subatomic particles of the universe from the beginning to work that way through the entangled subatomic quanta.

In molecular biology, it is the DNA emergence and changing process that is ingrained into the functions of life. It is the physics of the universe emerging and evolving matter and energy combining the elements which are also ingrained into the design of the universe that begins the entire process.

Hawking and Hertog concluded that the universe as a whole is more like a biological organism than we once thought because of the oneness of the whole entangled, interconnectedness of everything; and because the universe and it's laws of physics all came into existence through a birth and growth process — like all living things.

There have been many examples of scientists and philosophers who conducted rational scientific investigations which led to Deism, assuming that the natural world had been designed by a highly advanced, rational, and impersonal mind. Plato, Cicero, Maimonides, Kepler, Newton, and Albert

Einstein. They observed how the gravitational laws of physics formed the stars, planets, galaxies and cosmic debris. They measured the precise position of the orbits of planets and comets in our Solar System and how our galaxy formed because of completely natural phenomena. They couldn't help notice the biofriendly results of cosmology and astrobiology that made the emergence and evolution of life possible, wherever the conditions for habitation fall into place.

However, as you can see, Intelligent Design does not just claim that the natural world was designed by a rational mind, they insist on equating "design" to being guided by its Designer. Christian apologists and ID proponents claim that the fairly sudden appearance of the first complex animal forms in the Cambrian Period fossil record pointed to an "intelligent power" and attested to actions taken by a Designer, of "acts of a mind" ~ *Stephen Meyer.*

In their ID theory, they want to believe some very selective and suggestive evidence of a higher power is guiding the laws of Physics. That evidence actually points to processes that operate independently, not a guided and directed process. The truth will not promote anyone's atheist or theist selective, suggestive, and subjective hypothesis. The truth will wake you up with a jolt of awareness of an objective truth. The objective truth of dealing with facts as perceived without the distortion of personal feelings, prejudices, or interpretations; limited to fixed alternatives by reducing subjective factors to an absolute minimum. In other words, an obvious truth that can't be denied.

They quote the 1897 Oxford scholar *F.C.S. Schiller*, "it will not be possible to rule out the supposition that the process of evolution may be guided by an intelligent design." (Thus, the origin of that expression).

His use of the word guided is incorrect. It is a suggestive comment about the interpretation of the functions of the design by saying that the design is being guided along by a mind. The term calling him a "scholar" reflects the likelihood of a theist, a believer in the God of religion. So, that would mean that if we change the word "guided" to invented "by a mind," it means something that is not necessarily being guided, directed, intervened in, coerced, manipulated, or in any way divinely inspired by that mysterious mind of a Designer.

By saying that the design is be guided supports the delusions of Faith. This assumption by people that their beliefs in a god gives them authority over others turns people against each other in some very mind-numbingly inhuman and deadly ways. It was the major cause of wars, genocide, prejudices, and violence ever since we obtained or developed self-awareness of consciousness and continues unabated even to today.

20

Fine-Tuned?

When they say “fine-tuned,” as a “prior intelligent cause” it describes actions being taken by a god figure, adjustments made deliberately during the functioning of what was actually designed to function just fine without adjustments. They have to assume that their God performed the fine-tuning in order to justify religious beliefs. If you could build the perfect machine, you have no need to adjust it. Fine-tuned is their way of injecting into the definition of “intelligent design” a continuous tuning action being carried out by a god/God; a god being directly involved with the function of the design.

That is a continuation of the ancient belief system of many gods and goddesses that control everything. In this case it is one god with an entourage of angels, saints, and a son who was tortured and killed as a blood-sacrifice to his father, only to become a ghost who is apparently full of holes. Make it make sense? You can’t.

Our space-time bubble universe is a temporary living organism that will eventually die by burning itself out, following its functions to the inevitable end. It is not being adjusted by its Designer during this process. That is the wishful thinking of Christian apologists who have no argument, no evidence, and throughout the masses of sheep who enslave themselves to the fantasy and fiction of foolish "Faith," no clue.

The term "Mother Earth" has a deeper meaning than the average person realizes. God did not make us. No divine being exists and nothing was created by fiat. We were made by this planet, the stars, this galaxy, this universe, and all of its physics. It was intentionally designed from the start to evolve into a biofriendly cosmos about 13.8 billion years ago (at last calculation). Even though it is a form of projection, it is not one that is being manipulated by its Designer, but instead, being influenced by the existence of our entire collective consciousnesses, even down to the subatomic level.

If the process of DNA, the very building blocks of life, is not a random selection mutation process of order from chaos, (it isn't) and I doubt that it is some god of a religion sitting at a computer keyboard arranging DNA, (it isn't) then what is it? Some people mistakenly assume that the DNA process is too complex to be just a random occurrence. However, it is not the complexity of the process that is the issue. It's how and why the process is irreducibly complex and all working in unison. It is actually interconnected at the smallest level possible.

Random mutations are completely insufficient to explain exactly how the physiological changes occurred in the DNA advancements of living things. We must continue to question it with logic and common sense in humility so that we are ready to take the next steps in our own evolution when we awaken to the benevolent nature of the designer. All living things on Earth are rare in the universe and precious to whoever and whatever designed the process. We haven't found any kind of life outside of this planet. That should tell you how rare it is and how much we should protect all of it for its own sake. Science does not count magic or divine intervention as credible evidence.

> *Bart D. Ehrman:* "Faith in a miracle is a matter of faith, not objectively established knowledge."

Now that's hitting the nail on the head, telling it like it really is. That's what Bart does. Read some of his books and you will see.

This new theoretical hypothesis reveals that no matter how complex a function of life and the universe seems to us, it doesn't mean that it's magic or random occurrences. It manages to function the way it does based on its design, intended to work naturally on its own for a purpose that we have been denied knowing about to this point in the process of evolution.

21

Stephen Hawking

"*On The Origin of Time: Stephen Hawking's Final Theory*" {2023 Bantam} by Thomas Hertog was published just over two years ago now and it seems offhand like they might have agreed with me to some extent on this deist designed evolution theory.

This amazing book was reviewed by, Robin McKie, Science Editor, in an interview with Hawking's collaborator and author of the book, Thomas Hertog.

The book features Stephen Hawking's final thoughts on theoretical physics. From an interview with Thomas Hertog by Robin Mckie, Science Editor of *The Guardian*, March 19, 2023:

> "The problem for Hawking was his struggle to understand how the universe could have created conditions so perfectly hospitable to life," says Hertog, a cosmologist currently based at KU Leuven University in Belgium. "Stephen told me [that] he now thought he had been wrong and so he and I worked, shoulder to shoulder, for the next twenty years to develop a new theory of the cosmos, one that could better account for the emergence of life." Hertog said. "It leads to a new philosophy of physics that rejects the

> idea that the universe is a machine governed by unconditional laws with a prior existence, and replaces it with a view of the universe as a kind of self-organizing entity in which all sorts of emergent patterns appear, the most general of which we call the laws of physics… In the end, we both came to think of physics in a way much more like how we think of biology. We have put physics and biology on the same footing."

Hawking himself proclaimed, "I have changed my mind. My book, A Brief History of Time, is written from the wrong perspective."

> Robin Mckie, Science Editor, "Examples of these life supporting conditions include the delicate balance that exists between particle forces that allow chemistry and complex molecules to exist. In addition, the fact there are only three dimensions of space permits stable solar systems to evolve and provide homes for living creatures. Without these properties, the universe would not have produced life as we know it, some cosmologists argue it. According to Hertog, On the Origin of Time deals with questions about our place in the universe and what makes our universe fit for life. 'These questions were always in the background in our scientific publications.'"

In my summation, the design of the universe reveals an underlying force as powerful as gravity and just as ingrained in the universe to emerge life naturally and evolving it through beneficial changes in physiology into advanced states of corporeal beings with sentient and sapient cognition; wherever the conditions happen to fall into the right place at the right time. In Astrobiology, it is not disputed among scientists today that life can appear anywhere in the universe where all of a wide variety of conditions come perfectly together for it to happen. It's a purely natural function of the universe and it is my strong opinion that it is the direct result of a designed design by a MIND of some kind.

The Greek Philosopher Aristotle, centuries before Christ, was the first to say that the "Earth" is a giant rock floating freely in empty space. Religion said, "No, it's not." And that was the Law back then. Whatever the reigning religious authority proclaimed about our existence was the law. The Astronomer Nicolaus Copernicus said that the planets orbit around the Sun, the "earth" is a planet orbiting the Sun and rotating daily on its own axis. Religion said, "No, you're wrong." That was the law. Galileo said that the Moon, Jupiter's moons, Venus, and sunspots supported the idea that the Sun, not the Earth, was the center of the Universe. The Catholic Church said "No" and ordered Galileo to stop discussing ideas that conflicted with the teachings of the Church. Ideas that conflict with religious teachings are named heresy. Heresy of any kind was against the law and punishable by imprisonment or death. They did not kill him but ordered him to shut up. However, they did have other people who disputed them on rational grounds killed ever since the Holy Roman Empire's Catholic Christianity began.

A "heresy" translates to "against correct teaching," which of course, means not teaching what they want taught. My country has always been one demagogue away from that type of authoritarian and theocratic government. IT'S HAPPENING EVEN NOW. You may have heard the term "heretic." That's what is meant by that. You are a heretic if you are bold enough to confront their self-anointed judges and leaders about the lies and deceptions of religion.

Physicists for a century viewed the laws of physics as the lone absolute force in the natural world. The hope was that a powerful mathematical principle, i.e., "equation," would explain the {so far} unexplained.

However, Hertog and Hawking injects reason into the conversation,

> "even if such a powerful mathematical principle were ever found, it would hardly elucidate [make clear, explain] why the universe happens to be so remarkably biofriendly."

However, Thomas Hertog wrote, a

> "delicate balance of cosmic forces in which any slight change in 'basic principle properties' which are determined by the properties of elementary particles" and the forces acting in between them "would threaten any habitability in the universe from atoms and molecules to gravity itself."

To my mind he is saying that the biofriendly status of the universe had to be very specific and perfectly balanced a certain way for it to be biofriendly.

Hertog also mentions how a "celebrated writer and theoretical physicist," Paul Davies, described the universe's evolution random theory as the "Goldilocks factor." Goldilocks was looking for that "just right" porridge, chair, and bed (in the nursery rhyme) that ended up being the baby bear's things.

On the other hand, the design to inhabit and support life in the universe is very much like the Goldilocks factor. Life itself can only emerge on just the right planets, in just the right environments, star systems, galaxies and all of the necessary elements in the exact proportions to emerge and sustain life. Our planet just happened to be that right porridge, chair, and bed for the Goldilocks of life to emerge and evolve all the way to us and hopefully beyond. It's no happy accident, like Darwinists say, that life emerged and evolved here.

It is a fortunate product of its design that everything needed to emerge, maintain, and evolve life into complex organisms happened here — but accident? No. That is stretching the bounds of credulity. That promotes a worldview that the universe is *not* biofriendly or designed. That ignores the fact of the universe being a thing that doesn't have the ability to question anything, creating from itself material beings who can question their own existence. That isn't logical unless there is a MIND behind its design.

A mind behind the design has designed this universe to function toward a goal for a purpose that we do not know about collectively, nor even involved directly with the process of the purpose.

When we transitioned to walking, talking, and thinking primates, we knew what to call things; what to call each other, and how to improve on our language skills to form complex methods of communication. We began a journey of figuring out how we could use our surroundings to our immediate benefit and that led further down our path of evolving into questioning our existence. We became a level of sentient and sapient life unlike any other living thing. We are the product of the biofriendly universe itself. We evolved from primates, regardless of what you believe.

And that's exactly what we are — as proven by the history of changing over time until some of our primate ancestors began walking upright all the time and eating lots of meat, which provided them the nutritious fats and proteins that they needed to improve on cognition with their hunting and fishing skills, followed by social and communication skills. That is the key to explaining our evolution logically. Just being able to put names on things with articulate expressions of communication and language skills made hunting parties and building social structures more successful and set us apart from every other living thing.

22

Neil deGrasse Tyson

Neil deGrasse Tyson:

> "Objective truths are things that are true whether or not you believe it. And the methods and tools of science are uniquely conceived to seek out and establish objective truths. Then you have personal truths, things that you hold dearly. The problem with personal truths is that you can't convince someone else of your personal truths without some act of persuasion and in the limit, an act of violence. Then there are political truths. A political truth is something that is true because it has been incessantly repeated. And then you just believe it."
>
> "If your personal beliefs deny what's objectively true about the world, then they're more accurately called personal delusions."

Stephen Hawking, (From an interview with Neil deGrasse Tyson):

> "Collision with an asteroid is now one of the least of the dangers we now face. The last collision with an asteroid was 70 million years ago. Nuclear War, Climate Change and pollution are now much greater

> threats. Of course, it would be impossible to move the population of the Earth to a new planet. But if we can establish colonies in space, it would ensure the survival of the human race in the event of a disaster on Earth. Leaving Earth will also give us a new perspective and cause us to look outwards rather than inwards. It would also unite us to face a common challenge."

What are the cosmological implications of this conclusion? *On the Origin of Time* Thomas Hertog:

> "After *Maldacena's* discovery of the holographic nature of anti-de Sitter space, theories were quick to speculate that our expanding universe too might be a hologram. In the notebooks in which I transcribed some of my conversations with Stephen, I find musings about a possible surface description of expanding de Sitter space as far back as February 1999. But it wasn't until we had our top-down approach firmly on track, more than ten years later, that we began to pursue the idea of holographic cosmology in earnest.
>
> Unfortunately, by then Stephen was losing the slight control over his muscles that, miraculously, he had managed to retain for so many years of ALS."

23

Evolutionary Advancements

The humanoid skull advanced in size over a 3 million-year period from 500 cc to our average Homo-sapiens brain volume of about 1,400 cc.

It can be said that the brain and DNA of a younger generation, not just in humans, can contain a very slight but significant volume over its parents.

Richard Dawkins, The Blind Watchmaker:

> "To provide each succeeding generation with a significant survival advantage compared with the previous generation"

If that was true for the past evolution of the species, it means that the each generation of humanity are increasing in "significant survival advantages." When we add to that the increases in technology, we can see the significant differences between us now and them from the past.

All life that existed prior to 600 million years ago had only "soft bodies: no shells or bones to fossilize." After that period, we see the evolution of life here moving in sudden and intense changes — all similar in purpose, timing, and results — and could actually reflect sudden changes in a single generation of many species.

However, the overall reason for changes over time for all species is a drive, an underlying force, part of the very fabric of the universe itself, locked up tightly in all DNA, to emerge and evolve living things in increasing complexity in survival strategy, setting the stage for emerging and evolving life to it's ultimate design. This natural process will reflect itself in how the future of life will unfold as we go forward.

Jonathon Wells PhD, had a book out a couple decades ago called, "*The Politically Incorrect Guide to: Darwinism and Intelligent Design.*" It has deceptive statements and outright lies made throughout the book in favor of the religious side of design theory, intelligent design. (notice how propaganda works).

On page two "*Darwinism*" is explained fairly well:

> "Darwinism consists of the following claims: (1) all living things are modified descendants of a common ancestor; (2) the principal mechanism of modification has been natural selection acting on undirected variations that originate in DNA mutations; and (3) unguided processes are sufficient to explain all features of living things — so whatever may appear to be design is just an illusion." (pg.7) "The word "intelligent" emphasizes that "design" in this case is not just a pattern, but a pattern produced by a mind that conceives and executes a plan. Any natural causes involved are guided by intelligence. Writing a

> meaningful paragraph on a computer depends on various physiological, mechanical, and electronic processes, but without a mind directing them they would not produce the paragraph."

The design of the universe is not being guided or directed from anywhere. That is religious propaganda. A designed design that is engineered beforehand to operate on its own is independent of the MIND behind the design. Notice the use of the words guided, unguided and undirected in the above passage.

> (pg.11) "Another source of confusion in the controversy is that intelligent design is often mis-defined. The most common definition of ID in the news media is that some aspects of nature are so complex they must have been designed. Wrong definitions such as this may be simply due to misunderstanding, but some Darwinists deliberately mis-define ID in order to discredit it."

That much is very true! I have noticed how an atheist concept of design theory purposely discredits it by displaying the wrong definition of what design means when we say that living cells cannot change randomly without dying. We call that "Irreducible Complexity."

However, calling the process "unguided" and "undirected" implies to the reader that their design theory proposes it was, in fact, guided and directed by its Designer. A Designer who has never made direct contact with humankind regardless of what your ancient writings say. And why *would* IT? We were animals that evolved a conscious self-awareness of intelligent thought processes. It would be similar to us trying to make contact with an ant. If anything, a Designer is waiting patiently for us to evolve out of the natural selection mindset of animal psychology reflected in our capitalist economic system. Simply put, capitalism causes many people to have a bad life so that a few can live "the good life." The very basis of what a real Designer would consider "evil" is greed and selfishness. That which religion proclaims is evil is a distraction from their main motivation; money and power.

ID theory claims that the best answer to the great complexities of things like DNA, the inner workings of living cells, and the physics of the universe in general is one of miraculous guidance by a supernatural God or god-like entity with divine magic. Darwinism scoffs at that dumb argument, as well they should, and then proposes their own illogical argument in reply.

Per *Open Learn* (open.edu):

> "The four propositions underlying Darwin's theory of evolution through natural selection are: (1) more individuals than can survive; (2) there is therefore a struggle for existence; (3) individuals within a species show variation; and (4) offspring tend to inherit their parents' characters."

In other words, the strong survive better, which produces more offspring while the weak do not. So, surviving species pass their genes on to their next generation with any physical changes to them that were necessary to help survive their extinction. It's in the process that leaves the question of how the physiological changes were made in our DNA, not that the survivors continue to procreate and flourish. There is a biological phenomenon which takes place in all species that changes their structure for enhanced survival, but it isn't driven by randomness or magic — it is a designed function in the biofriendly processes of the universe itself, like entanglement is.

There's no disputing that evolution happened by species of life changing their bodily structure and composition to meet the challenges that the world throws at them. That, by itself, doesn't answer how the physiological mutations took place in evolving species, it only proposes the why it happens. Every step of our evolution has been accompanied by slight changes in our physiology which makes surviving easier. The average height of men in 1920 was 171.4 cm, then during WW2 was 174.77 cm. In my birth year 1956 it was 176.27 cm. In recent years this has fluctuated. During the Civil War, the average American male was only 5 feet 7 inches compared to today's average of 5.9 ft.

Why would that be happening because of natural selection? It doesn't appear random. It does appear to be part of the design of everything to get bigger, faster, and smarter under the pressure to survive as time goes by.

24

The Sixth Mass Extinction

Rare Earth, pg.180:

> "any intelligent species co-opting a planet's resources in the service of advanced technology and agriculture will necessarily cause a planetary mass extinction."

"*In Brief: Winter 2007-08*" from "*EarthJustice*; *Because the earth needs a good lawyer*" reveals a few things. From a small section on page 13 of this 29-page booklet I learned of the dangers of phosphate mining...

> "A Phosphate Mine Poisons Idaho Waters", "In Southeast Idaho, in and around the Caribou-Targhee National Forest, some 17 phosphate mines have been designated as Superfund sites {designated as needing super-cleanup effort} as a result of serious contamination. Current and past phosphate mining in the region has resulted in ongoing widespread selenium contamination of streams, groundwater, vegetation, and soil... [Selenium can trigger serious toxic effects in fish, livestock, wildlife, and humans. Reproductive failure, severe deformations both at birth and in adults, and even death can occur. Once released into our land and water, toxic levels of selenium persist for hundreds of years, threatening the health and prosperity of future

> generations. Selenium also builds up in the food chain with increasing concentrations in plants, fish, and animals — including trout, deer, and elk that are eaten by local hunters and anglers and their families."

Human capitalist enterprise is the major, if not exclusive, cause of planet-wide mass extinctions and toxic pollutants. Apparently, humans are the destroyers of the earth as well as their future existence.

"*The Sixth Mass Extinction: An Unnatural History*" by Elizabeth Kolbert:

> "(In) an extinction event of our making, what happens to us? One possibility — the possibility implied by the Hall of Biodiversity — is that we, too, will eventually be undone by our transformation of the ecological landscape."

The logic behind this way of thinking runs as follows:

> "having freed us from the constraints of evolution, humans nevertheless remain dependent on the Earth's biological and geochemical systems. By disrupting these systems — cutting down tropical rain-forests, altering the composition of the atmosphere, acidifying the oceans — we're putting our own survival in danger. Among the many lessons that emerge from the geological record, the most sobering is that in life, as in mutual funds, past performance is no guarantee of future results. When a mass extinction occurs, it takes out the weak and lays low the strong."

The Anthropologist Richard Leakey has warned that;

"Homo sapiens might not only be the agent of the sixth mass extinction, but also risks being one of its victims."

On the bright side, humans have a relatively long history of finding solutions to problems or simply improvising, adapting, and overcoming their adversity. However, I doubt that the damage we've done to Mother Earth can be reversed with centuries of lying dormant, or humanity adapting here to what is coming, or that we can travel to and make homes on other planets, moons, or wherever.

Elizabeth Kolbert goes on to proclaim that this mass extinction event we are now fast approaching may turn out to be our undoing, but it will be our "most enduring legacy" either way. With a sour note she concludes;

> "*The Sixth Extinction* will continue to determine the course of life long after everything people have written and painted and built has been ground into dust and giant rats have — or have not — inherited the earth."

Darwinism claims that evolution begins with biological organisms mutating randomly. However, the pattern we see in the evolutionary changes of birds on the Galapagos islands reveals a knack, a driving force, an inclination in nature, to make beneficial physiological changes in species rather than random mutations. It is not a random process if there is a clear intention in nature to modify the physiology of living things to survive better. It was the purpose in evolution to advance life from early simple forms all the way up to large dinosaurs and then thousands of years later to walking upright primates.

That is not random at all. The birds on the islands had to improve on the design of their beaks, which took place over time, toward the changes that helped the species to eat, survive, and thrive. These changes were inherited by their offspring on and on until most of the species had specific kinds of beaks for eating, depending on what they were eating on which isolated island they lived on. Those who did not have improvements in their ability to survive, usually died off. In fact, the driving force behind the emergence and evolution of life is a naturally occurring process to enhance the survival of the species by advancing beneficial complex forms forward.

This can be attributed to the evolution of all species since the first. It is not an emergence and evolution by random occurrence or by magical fiat. In this way, our universe itself was designed to create itself as well as life and evolve it all from the beginning. I believe this held much weight with Stephen Hawking's decision to proclaim the universe as "biofriendly" and his "top-down" cosmology. If it appears that the universe has been working toward making a home for living things all along and as the evolution process proceeds forward into the inevitable future, the results will reveal the reason for the intention behind making it biofriendly. Whatever that may be, the future will reveal it whether we are still around to see it or not.

Hawking and Hertog established that our quest for understanding the universe could inevitably involve metaphysics to some degree.

(*Glossary Definition: "Metaphysics" – PBS*)

> "Metaphysics" "the study of nature of the human mind, the definition and meaning of existence, or the nature of space, time, and/or causality."

"Metaphysics" definition from *Webster's Ninth New Collegiate Dictionary* (1983);

> "a division of philosophy that is concerned with the fundamental nature of reality and being and that includes ontology, cosmology, epistemology."
>
> "Epistemology" "to understand, to know" ": the study of a theory of the nature and grounds of knowledge esp. with reference to its limits and validity —"

Before science began to know what this planet is, the stars, moons and galaxies; no human knew what it was or even had a clue. They called the planets gods and named them after Greek and Roman mythological gods. Judaism thought the stars were angels, the heavenly host. Since we are the universe looking around at itself, we can now see it for what it really is and right now we are beginning to see that it is a designed design by a MIND of some kind because of the universe's biofriendly intentions. The only way to make that fact logical is by realizing that a MIND of some kind is behind that intention.

To the old school of ancient science and religion, that's not what they believe or want to accept. Science decided long ago that there isn't any reason or purpose to the existence of the universe. Even those religious intellectuals who believe in the existence of a guiding hand in the making and functioning of the universe do not ask the question of why the universe and everything in it exists. They are all still stumbling around in the dark as part of the design. They just aren't questioning anymore about where the sun goes at night. That much we can know for sure now, thanks to telescopes and astronomers.

Why though? Why did a MIND of some kind design, engineer and implement this amazingly biofriendly universe? That is the question that has been the most elusive of all to the intelligent design advocates who look to the heavens for their answers. The "why" might be obtainable now with just common-sense reasoning and logic. Just not by divine revelation or direct biblical claims in some infallible scriptures written by human beings ages ago.

Evolution is the number one reason that everything exists. Whether you like it or not, life has evolved by naturally designed processes that was conceived and invented by a creative and ingenious MIND of some kind to emerge and evolve thinking material living organisms into existence from nothing and grow them (us) into complexity for a reason we cannot be sure of at this time. However, we can make an educated guess, now that quantum physics is better understood and a lost religion is being uncovered in front of our eyes that both combine to reveal a possible answer based on scientific observations of subatomic particles with the religious belief of a general resurrection from the dead at the end of human history, whenever, however that happens.

The universe's supreme gift to us is our refined ability to question everything with cognitive critical thinking skills. Not knowing why the universe did that, we can assume that there is a secret knowledge. We can also assume that neither religion nor science knows what that knowledge is. Religion expects you to abandon your critical thinking questions and replace it with a faith in illogical beliefs where there is no hint of solid answers. Science went in the opposite direction long ago but got hung up on devout atheism which blinded them almost as thoroughly as faith does. Once they were satisfied that their answers were enough to refute the other side they resist questioning their own theories.

An "*Impersonal Universal Consciousness*" (Geoffrey Riggs) is the best description that I have heard of the MIND behind the design.

The possibilities of extraterrestrial biology have grown increasingly plausible. An enormous body of science is being assembled to analyze and explain the origins, characteristics, and possible extraterrestrial dimensions of life. NASA describes "Astrobiology" as "the study of life here and beyond."

The biology of complex lifeforms initially began in the oceans and adapted to living on land. Those first forms of life here were aquatic when some began to develop ways to absorb oxygen through breathing with an early form of lungs, besides still having and using gills. So, when they needed to escape predators in the water they could crawl up on to the land. This drastic act of survival not only began but enhanced their ability over time to breathe air. The driving force in the evolution of the species is the drive to survive. Not random or miraculous mutations as others insist. They began to develop further into land creatures until becoming strictly land animals and losing their gills in the process. Whales are somewhat of a mystery because there appears to be evidence that they lived as aquatic creatures first, became "terrestrial," living on land, and then going back to the water and becoming the whales we know today.

When small dinosaurs survived the asteroid impact that killed off all of the large species, they eventually grew wings, also to ensure that they could escape predators and expand their territorial domain, and henceforth, they became all of the birds.

Lifeforms did not think, "Hey, I need wings so I can fly, or I will be eaten." Or "I need to live on the land to get away from the predators in the ocean, I need lungs." But that is effectively what happened without them knowing it. Our apelike ancestors saw a great need to stop living in trees like other primates and started living on the ground in caves and debris huts of piled up thorny branches to protect themselves and their young. This happened somewhere between 6 and 4 million years ago. From that point in history, we saw the need to walk upright on just our legs.

Once we began eating meat exclusively, meat protein increased our brain size and mental capacity. Our social skills, communication skills and hunting skills enabled larger and larger groups of our species of humanoid to start working together for the benefit of the group, even to develop the first economic system and civilization, which they saw as beneficial for everyone rather than fighting over resources.

That's right, *socialism*. They noticed how wolf packs worked together in attacking large prey and traveled together in a procession of a leader, a few of the stronger wolves following behind him, and the rest of the pack behind them with one or two strong wolves taking up the rear for defending the young. The very old were somewhat isolated behind everyone else, being a favorite target of predators, to protect the group. Life and death was part of the facts of life that couldn't be totally avoided.

The survival of the species of life is an inherited force within all life from the beginning of our planet's history as soon as conditions here allowed for the first single-celled organisms to emerge, survive, thrive, and procreate species almost 4 billion years ago. That driving force to live is not a trait that came about randomly, nor was it an act of God adjusting the design. It was/is an integral part of the functions of life and the universe as it evolved from the Big Bang forward over the long expanse of time that our universe has existed. The universe is all connected for the purpose that it has employed for making itself biofriendly. To religion, that prospect negates their profit-making guided and directed authoritarian theism and to science, it negates their random order from chaos equation.

25

Astrobiology

Wikipedia defines "Astrobiology" as; "the multidisciplinary field that investigates the possibility of life on other worlds." For me, it is another important steppingstone to answer this big question. An integral piece of the jigsaw puzzle that I endeavored to solve. The one question that neither science nor religion has a logical answer for, *why*?

In the book, "*Rare Earth: Why Complex Life is Uncommon in the Universe*" (2000) By Peter D. Ward and Donald Brownlee, we can take a look at some of the pivotal moments in evolution. The fall of the dinosaurs occurred 65 million years ago by an asteroid impact and the following conditions that it created. The avian species evolved into birds from small dinosaurs and pterodactyls. There have been roughly 700 species of animals discovered and named that go back to after that extinction event occurred. There could be another planet like this in some other galaxy, but it must be the right size, the right orbit of a suitable star at the correct distance from it with very many other supporting features and conditions to have including all the elements, atmosphere, oxygen, hydrogen, etc. to support life as we know it through DNA. However, it took a huge amount of time for the process to get to creating us. That was about a third of the life of the universe.

The variations that occurred here to make it possible for life to appear/emerge and evolve all this time are astounding. DNA is defined as "Deoxyribonucleic acid" (*Britannica*) and is an organic chemical that holds genetic information and instructions for protein synthesis. It is found in the cells of every living organism on Earth and according to astrobiology, the only kind of life in this universe that is possible is from DNA.

The earliest evidence of life comes from simple-celled, biogenic carbon signatures and stromatolite fossils discovered existing 3.7 billion years ago in meta-sedimentary rocks from western Greenland. Most astrobiologists round out that number at 3.8 billion years ago that life first appears to have began. The formation of the Earth and our solar system occurred about 4.3 billion years ago; give or take an eon or two.

In 2015, possible remains of biotic life were found in western Australia. The earliest known life forms, according to another source, are believed to be fossilized microorganisms found in hydrothermal vents considered to be about 3.42 billion years old.

The likelihood of a Solar System exactly like ours being out there somewhere is more remote than you might think, in spite of how enormously huge the universe is. It is how uniquely perfect everything had to be here to emerge life and evolve into us. It all fell perfectly into place over a 4 billion-year period to produce us and 3.5 billion years of that alone, just to evolve life from the original simple forms into complex multicellular forms.

This planet sufficiently emerged organic molecules that have evolved over billions of years into the sapient and sentient life-form that we are today and hopefully will continue evolving well into the future.

Locked somehow into all DNA is that survival driven metamorphic process to enhance all physiology through genetic upgrades, an evolution of complexity in all life. The evolution of the universe, including the evolution of life, are the keys to understanding a truly logical purpose behind this designed design by a MIND of some kind. Hence, we see all the physical advances of simple lifeforms over time into complex forms. Hence the arms of small dinosaurs forming through metamorphosis into the wings of birds, up to the advances made by early hominid bi-pedal primates into us over about 5 million years... a drop in the bucket of time compared to other species' paths and then finally into complex cognitive awareness of consciousness (whenever that happened).

That driving force in life to survive, to "Never Quit" (Mykel Hawke), is very much associated with the process of evolution since the beginning. We know that. It should not be in dispute. What is in dispute, of course, is how that process is either a random chance occurrence or intervening actions by a deity and no other reason is allowed into the conversation by devotees of atheism and theology. However, what those in science and religion are reluctant to consider is that the process is interwoven into DNA as part of the fabric of all life functioning naturally like the huge interconnected living organism that it is. That will be the direction that theoretical physics will go. My theory must be recognized as legitimate. The scientific and religious communities should be aware of this. Stephen Hawking and Thomas Hertog were realizing it and trying to make sense of it before Stephen could no longer communicate at all.

The benevolence of an unselfish mindset, geared towards service to others without payment or applause, is the only way to avoid our own extinction. It's about feeling obligated as a living, breathing human being to help the cold get warm, the blind to see, the crippled to walk, the hungry to eat, the thirsty to drink —THAT is the essence of what a consciously self-aware spirit must achieve to begin to qualify for something beyond life. The Designer of the Universe, the Impersonal Universal Consciousness, won't allow any Adam-and-Eve type of arrogantly selfish individuals have access to the holographic program that the universe is. That wouldn't make any sense.

With 2 trillion galaxies in the known universe, it should be considered an absolute fact that somewhere out there is another us. Therefore, if we end up destroying ourselves before developing the benevolent nature that a Designer has intended [let us make man benevolent] for us to evolve into, it won't mean that the purpose of the universe has failed completely with its task. It would just mean that it failed to happen here.

26

My Pinball Machine Analogy

Many stars called “spiral stars” or spiral arms, are the result of newly born stars emerging in clusters before separating from their siblings which could create dangers for solar systems like ours if one was in our galaxy. For 4-billion years, just over a third the history of the universe, had to be stable in many ways for this Milky Way galaxy to emerge and evolve us. Most other galaxies are not shaped in a way that would have conditions favorable for life. However, this is one huge universe. It just means that any intelligent life in our universe would be remarkably similar to us in many ways but actually be very few and far between. The most advanced forms of life out there are watching us, if that, without getting involved. If they can see how we are slowly but willingly destroying everything here to profit a few in the short-term, you can see why ET stays away.

The dangers to our solar system include “supernovae” and close shaves with other stars and interstellar clouds. Those who look at the fossil record of life are matching up some of history’s mass extinction events with our solar system’s history. A history of revolving around not just our place in the solar system but our solar system revolving around our place in the Milky Way Galaxy. It takes our little solar system between 225 million to 250 million years to completely revolve around the Black Hole in the middle of our galaxy.

We know how fragile life is.

The basis of our existence begins and lies in the stars which creates all forms of matter, then dispersing that matter when the star dies or goes supernova. This throws the elements out into the galaxies to form planets, moons, asteroids, meteors, and other cosmic debris. Meteorites have iron and nickel with small amounts of sulfone and carbide minerals. As our planet revolves around that specific star we call the sun at a comfortable distance to support life, our whole solar system must revolve around in its galaxy at a comfortable distance from its Black Hole in the center.

Here is my "Pinball Machine Analogy" for explaining how random occurrence is part of the design that functions our existence. I am not familiar with any already established analogy exactly like this. The old-fashioned pinball machines that I loved to play growing up are an example of how randomness can be designed into the holographic program itself. You start by sending a metal ball up to the top of the machine which tilts downward. There are "flippers" in various places that you can strike the ball with as it rolls down, thus keeping the ball in play, and with two specific flippers near the bottom (one on both sides) of the machine. As the ball rolls downward, it hits "bumpers" and falls into slots that score points. The goal of the game is to keep the ball in play as long as you can to amass more points. The ball bounces around in semi-controlled patterns, maneuvered to an extent by the person working the flippers to keep scoring points simply by keeping the ball in play. It is dependent upon the random occurrences of where the ball will go that was designed to provide the player with some power to manipulate the ball until it goes down the middle and out of play. We live in a random occurrence existence that we are designed to manipulate to a certain extent for our survival.

75,000 years ago, the Toba super-volcano caused a "bottleneck population event" that killed off almost all humanoid species around the world except for the southern half of eastern Africa, reducing the global population down to an undetermined number individuals. After that, human beings moved northward from there and then throughout the world, eventually creating the world you see now. After that super-volcano blew, humanity's numbers were squeezed down to what might have been a few tens of thousands from a mix of the last few species that included "Homo sapiens" (us) and "Neanderthal."

The Neanderthal species later became extinct while merging to some extent with Homo sapiens in the millennia that followed. I think that last calculation I heard is that everyone has at least 2% Neanderthal DNA with some people having slightly more. The world is today looking at the probability of a similar event recurring by purely human-caused reasons. That brings with it the possibility that our species of human will be nearly destroyed (if not completely so) as another kind of humanoid might take over in the thousands of years ahead, if not much longer. That is conceivable in logic to be a strong possibility at this point. There's also the possibility that present humanity comes to a complete and final end with no new species of humanoid following us to replenish the Earth.

We share 99% of our DNA with chimpanzees. No, we didn't evolve from them or from apes, we link up in the early family tree of all primates. We all evolved from common ancestors. We share 98% of our DNA with gorillas and orangutans. There should be no doubt that we evolved from primate species. These changes in species taking place over time in our DNA (the building blocks of life in our universe) have accumulated through mutations that were designed to advance life forward in complexity. It may only appear somewhat of a random process on the surface but at the deepest level it seems extremely logical to be part of the interconnected nature of the universe itself. This was all designed to function as it has. Instead of saying, "shit happens for a reason," we should say, "That's the way the ball bounces." I know I said that already, but the pin-ball machine analogy brought some clarity to it.

Since the dawn of the industrial age, the average world temperatures have increased by 2 degrees. Scientists have been saying for generations now that 2 degrees is the point of no return to a mass extinction event, not just from Climate Change alone but exacerbated by the world's other problems associated with man's misrule over his resources.

The melting of the world's glaciers and ice-caps in the polar regions will melt away completely, causing increasing desert takeovers in various places and extremely high sea-level rise. We are feeling the beginning of that now. Each of the years 2021, 22, and 23 were all record-breaking hot temperatures worldwide. By the year 2070, one/fifth (1/5th) of the planet will be dry desert. (PBS, Evolution Earth) As our younger generations begin to fully realize what the consequences of industrialization has caused to their future prospects, they're going to be quite angry at the industries, politicians, and religions that led them to this global catastrophe. I thought by now, the youngest of the generations maturing at this time would be voting in massive percentages for progressive politicians. They aren't. It is perplexing that so many people ignore the coming fate of the world.

It was the oil companies who could have prevented it if there had been a willingness long ago to try, as they have known full well for the last 80 years -at least- that this would happen eventually. However, it's happening much sooner than all previous scientific analysis on it expected. Every time a new study was conducted on Climate Change over the course of my 70 years of living, I noticed that the prior predictions were always well short of what actually happened. Was that intentional? Probably so, when the only way to conduct research on climate required financial backing from somewhere.

With the number of people on this planet over 8 billion, you will see increased pollution and hazardous waste in all the habitable regions and beyond.

The historical Jesus was more of a humanist-socialist in his sayings than the different forms of Christologies that mutated away from that original socialist Jewish sect. The historical Jesus was not an advocate of politics, war, family values, or obsessions over money! Just the opposite. We must embrace his mindset, or else, the extinction of many species will adversely effect the struggle to continue our own species.

If humanity is going to be spared from extinction it will involve the philosophy based on socialism called,

> "Humanism". (*Merriam/Webster's*) "a doctrine, attitude, or way of life centered on human interests or values; especially a philosophy that usually rejects super-naturalism and stresses an individual's dignity and worth and capacity for self-realization through reason."

Albert Einstein supported humanist organizations well into his latter years.

27

Rare Earth: Why Complex Life is Uncommon

In the theories of Creationism and intelligent design, Creationists believe the universe is permanent, supernatural, and miraculously guided. The terms "miraculous" and "guided" are used to define an action being taken, an intervention, not a function of a process that was designed to occur naturally. This is why they don't care about what they are doing to our civilization or the planet's ability to sustain life. They are depending on divine intervention and miraculous guidance from above so they can ease their conscience over what the effects of their actions are doing to the war-torn corners of the planet and to the environment, the climate, and in effect, all life on this planet.

"*Rare Earth: Why Complex Life Is Uncommon in the Universe*" by Peter D. Ward and Donald Brownlee; pg. 57;

> "There might be many kinds of life elsewhere in the Universe, and there is a great deal of speculation among scientists about whether DNA is the only molecule on which life can be based or one of many. It is certainly the only one capable of replication and evolution on Earth, and all life here contains DNA. The fact that all organisms on Earth share the same genetic

> code is the strongest evidence that all life here derives from one common ancestor."
>
> "Perhaps life may originate on any planet as soon as temperatures cool to the point where amino acids and proteins can form and adhere to one another through stable chemical bonds."
>
> (page 61) "Life seems to have appeared simultaneously with the cessation of the heavy bombardment (of cosmic debris). As soon as the rain of asteroids ceased and surface temperatures on Earth permanently fell below the boiling point of water, life seems to have appeared."

Any scientific theory must rely on detailed physical evidence, standard logic, and sometimes intricate detail. That's reality! "Evolution" is merely a process of life changing over time in complexity. The real question now is whether it occurred by random chance, divine intervention, or a designed process occurring naturally. Many people don't see that yet, but *that is the question*!

> *Rare Earth* pg. 140; "On Earth it is clear that the evolution of animals occurred not as a gradual process but as a series of long periods of little change, punctuated by great advances..."

Neil DeGrasse Tyson;

> "Life got started almost as fast as it possibly could have in the early Earth, after Earth cooled down from meteor impacts... Once it cooled down, complex

> molecules form within 100 million years — life appears... To suggest that we are the only life in the universe [14 billion years old with 2 trillion galaxies] would be inexcusably egocentric."

To find the answers, the devil is in the details. (even though the devil is not real)

When scientists mapped out the regulatory system of sea urchins it resembled the "complex circuitry of computers" (from the book, "*The Edge of Evolution*", by Michael J. Behe). "Living cells at the molecular level are 'irreducibly complex'. They cannot change their structure, components, or composition randomly."

So what is "irreducible complexity?" The living cell is like a mouse trap. If you take away any part of a mousetrap to replace that part, it will cease to function as soon as one part is unhooked. This is what is called "irreducible complexity." Its parts can't just change, or mutate, without interrupting its functions unless it changes simultaneously without hesitation in its function, allowing the function to continue unabated. If you must change a part of a mousetrap, no matter which part it is, the trap is not functional in between the exchange of its parts. Each of the various parts of the mousetrap have been designed to function in catching mice. Without every part continuously working, the mousetrap is dead. The cells of living organisms cannot change their physiology randomly, which is what Darwinist claims about evolution have suggested.

Living cells cannot mutate randomly or they cease to function, which means they die. If you want to change any part of the mousetrap to make it work better or to last longer, you can just take a part off and replace it with a more advanced part. A living cell functions on the same principle of many moving parts working together at the same time. If you try to take any of those parts off to replace it, the cell itself will die, ceasing to function, unless it changes the part somehow without interruption, like simultaneously. The discovery of the interconnectedness and entanglement in the atoms and the innermost components of quantum physics is the answer. Evolution is not a random process or divine intervention. The mousetrap underwent changes in its design by us until it functioned at peak efficiency. Living cells go through similar processes during evolution but function as one whole self-aware and connected system. As explained above, the subatomic particles of atoms function together as if one living entity.

Living cells function in the same way, yet Darwinism claims that it came about by mutating randomly until it stumbled into the correct size, shape, and composition needed to function in its particular environment for its particular purpose. Recent discoveries in the compositions and functions of living cells do not reflect processes of random mutations as their cause for changing over time. Instead, it reveals patterns of irreducible complexity and a driving force in nature and physics itself to make beneficial changes to life simultaneously within the biology of lifeforms. Moreover, it does not mean that the process is being directed, “guided” or intervened in by the

Designer. That claim by supporters of intelligent design is baseless. It is an apologist Christian strategy to claim evidence of a God who intervenes in human affairs to justify pretending that they are doing His work, obeying Him, worshiping Him, and judging others for Him. It is a huge mistake and a reflection on human arrogance. There is no Him! The processes of life are like the processes of Quantum physics. It could even be the cause. They are interconnected and entangled. In the case of the emergence and evolution of life, they are very tightly interwoven into the fabric of the holographic physical laws and astrobiology of the universe.

"*The Case for a Creator*", by Lee Strobel, (interview with a scientist):

> "Life is actually based on molecular machines... They haul cargo from one place in the cell to another; they turn circular switches on and off; they function as pulleys and cables; solar-powered machines capture the energy from light and store it in chemicals. Molecular machinery lets cells move, reproduce, and process food. In fact, every part of the cell's function is controlled by complex, highly calibrated machines."

What machine was invented or perfected by throwing random things together? Some machines are designed to require a "guiding hand" to work it, like a computer or the laptop that I'm guiding and directing letters into words in highly complex patterns of expression. The machine of life (or the living organism that the universe is) and the designed natural functions of the "physics" of the universe are not random things, it was designed by a MIND of some kind to function as it does completely naturally.

We know from the experts on the Earth's first formations of life that a pool of organic building blocks could produce life. Life from liquid water and organic molecules can rapidly develop into replicating organisms ("replicating" means increasing in numbers by procreating), beginning the process of evolution here. Once its information storage and transmission system {DNA} had been achieved, natural selection and the drive to survive pushed organisms to grow and become stronger, fitter and survive easier, when necessary, by gradually achieving greater complexity as it evolved — sometimes over very long periods of little changes interrupted by relatively short bursts of dramatic changes required to adapt to harsher conditions. The bulk of these adaptations happened in response to the mass-extinction events of varying sizes and effects from the officially established five mass extinction events prior to the one on our doorsteps.

It's considered a "quick start" in evolution that once all the elements and conditions here were survivable for living things, it happened relatively fast. That means that the prospect of life emerging elsewhere is very good, especially given the prevalence of organic molecules in nebula throughout the galaxy. However, when we try to discover exactly how amino acids and elemental molecules coalesce, we fall short.

> "Probiotic chemistry leaves no fossil record, making it difficult to piece together the chain of events."

("*Astrobiology: Exploring Life in the Universe*" by Lewis Dartnell)

Scientists have tried to recreate the process of emerging life from inorganic materials in controlled laboratory experiments, without success.

Unfortunately, exactly what happened on Earth between the ceasefire of heavy bombardment from cosmic debris when the planets were still forming and when the emergence of living cells occurred here are very much a mystery to science to this day. Just another one of those things that we might assume scientific methodology has discovered, has fallen short. It is still very much a mystery that we cannot replicate the process.

Roughly, 70% of the stars of our Milky Way galaxy are not likely to create life and 75% of those that could are not likely to successfully emerge and evolve life to any relative degree. Also, bigger and brighter stars than our sun and small red dwarf stars are places where the conditions and ingredients for life are not favorable at all for several reasons. Our solar system has rare qualities apart from others that we can observe with the latest technologically advanced space-telescopes. The average galaxy beyond ours is elliptical shaped and chaotic where ours is disk-shaped with spiral arms and is stable. It has an elliptical bulge in the center and a spherical halo that is denser closer to the Galaxy's center.

As far as we know through Einstein's proven theories, nothing made of matter can reach the speed of light or especially go beyond it. That technology could come in the future, but so far, the way it adds up, reaching the speed of light will be impossible for us. We will need to somehow bend space or create stable worm-holes that we can manipulate with to travel through. However, as we approach the speed of light, time itself slows down with it. It could theoretically kill a person one way or the other. It is therefore impossible at this time for us to travel to another galaxy or other star systems in our own galaxy. We are still uncertain that we can send people to other parts of our own solar system, even just to our closest other planet, Mars. That will come eventually, if we don't go extinct first. Because, humanity will run out of time for colonizing outer space before our planet becomes a wasteland. Time is of the essence.

28

Astronomer Carl Sagan

Carl Sagan spent a year of his life involved with the Mars Rover project which made a high-tech search for any signs of life on Mars. He had this to say about the experience;

> "There was nothing. Not a mouse, not a footprint, not a microbe, not even an organic molecule. The lesson is that you can have dozens of worlds, some of them with breathtaking splendor and beauty without life on any of them. Life is something rare and precious. There is something extraordinary about the planet that we are privileged to live on."

Generally speaking, science had assumed that the physics of the universe were set in place, as if set in stone, as early as the Big Bang and remained unchanged throughout its 13.8 billion year history. We know now, and Thomas Hertog's "*On the Origin of Time*" book can verify this, that the physical laws of the universe had to evolve in the same manner as life has from nothing into the biofriendly time-space continuum that it is today and will continue to change over time (evolve) until it all burns out.

> *The Primordial Universe*: "According to the Big Bang theory, the universe goes through many processes after the Big Bang. The Planck Age, the Inflation Period,

> and many other things that happened during the evolution of the universe. Each stage of the evolution of the universe has its own significance. Many great men have explained to us in detail."

Creationists have been claiming that the reason we didn't find any fossils of earlier forms before the Cambrian Period explosion of complex life is because all species were supernaturally created at that time, "in the beginning," as it says in Genesis, "according to its kind." Scholars have debunked that translation in the first chapter of Genesis that says, "according to its kind." Those verses should say, "and others of its kind" instead.

For the first 3.5 billion years (approx.) of our planet's existence, no life is detected in the actual fossil record because the life that was there were only soft-tissue species which evolved from early simple organisms; bacteria in the beginning, with no bone structure in which to leave a fossil. Modern methods of detecting them in the strata have discovered this. The "Prokaryotic" were single-celled organisms that consisted of a jelly-like structure and would not have left a fossil behind. There are also the Eukaryotic cells that began about 1.4 billion years ago. At one point, we thought that the Earth was created 1.4 billion years ago. We can detect that life was here well before the Cambrian now with advanced technology that showed us how old the Earth really is. Now we know that lifeforms changed dramatically from earlier forms after the Cambrian Period explosion of highly complex species, producing the forerunners of the dinosaurs as well as an incredibly wide variety of varied oceanic species. One of the amazing things about our growing knowledge of everything is that we learned it in stages. In antiquity, because of Christianity, the world thought that our planet was 7 to 10 thousand years old. Then our ability to look closer discovered that our planet was millions of years older than religion was saying, then over a billion years older, then almost 2 billion years older, and finally the nearly 4 billion years with the first sign of life forming 3.8 billion years ago. In the same sense, we have discovered that our planet is in a star system, then in a galaxy, then in a universe of galaxies. However, keep in mind that it may all be an extremely advanced holographic projection that we are seeing and calculating with.

NASA once made the statement, "There are at least 100 billion stars (in our galaxy), and on average, each star has at least one planet. This means there are potentially thousands of planetary systems like our solar system."

It's strange that NASA would make such a deceptive statement, implying that because there are so many stars in our huge galaxy with so many star systems, there are potentially "thousands of planetary systems like" ours. That's deceptive considering that plenty of accredited astronomers and astrobiologists know for a *fact* that the conditions within our solar system very much appears to be so uniquely suited for life that there would not be potentially hardly any others at all that are suited for life's appearance and evolution in our galaxy. The astounding physicist, Professor Brian Cox, says that it is conceivable that we are the only civilization in this galaxy with sapient, sentient, thinking, conscious beings.

That's a detail NASA failed to include. Instead, they imply that there could be thousands of other "planetary systems" just like our own.

Having the stable environment here and the right conditions, elements, physics, gravitational forces, and many other factors to provide habitation for living things is becoming increasingly unlikely in the universe, as our understanding grows. Yes, NASA was being deceptive, because they need public support and the best way to do that is to make statements that are encouraging space exploration and popular with the public who wish us to find another planet like ours... desperately... where we can have the hope that we might be able to move to one someday since we are trashing our own world.

> Per *Space.com*; "The Milky Way is a barred spiral galaxy around 13.6 billion years old with large pivoting arms stretching out across the cosmos. Our home galaxy's disk is about 100,000 light-years in diameter and just 1000 light-years thick, according to Las Cumbres Observatory. (Apr 18, 2023)."

29

Theoretical Physicist: Thomas Hertog

From Thomas Hertog's book "*On the Origin of Time: Stephen Hawking's Final Theory*" are some interesting points that they have considered in recent years;

> "The quantum cosmology that we propose reads like the universe's history from within and as one that includes, in its earliest stages, the genealogy of the physical laws. In our view it is not the laws as such but their ability to change that is fundamental. In this way top-down cosmology completes the conceptual revolution in our thinking about the universe that Lemaitre initiated. When tracing the universe back in time… a deeper level of evolution opens up in which laws of physics themselves change. Our discovery is a sort of meta-evolution, a stage where the rules and principles of physical evolution co-evolve with the universe they govern."

> "It is a striking property of top-down cosmology that it has a built-in mechanism that limits what we can say about the world… the cosmos guards us from wanting to know too much."

Exactly!

Part of what *is* understood from Hertog and Hawking's work is that the emergence and evolution of biological life and the emergence and evolution of our universe are two distinct parts of one "giant evolutionary tree." From high complexity and low complexity (looking at it back through time) — the "astrophysical, geological, and chemical" with its own languages shows a "universal wave function that weaves them all together." Life in the universe and the emergence of the universe itself are mutually beneficial.

> "the emergence and evolution of life and the emergence and evolution of the universe are deeply connected, representing two stages of a single overarching historical process… Such an extraordinary arch reveals a profound and powerful unity in nature."

See what that means? We've depended on science's determination to explain the existence of the universe with a specific mathematical equation. A solid discernible theory based only on empirical, definable, proven evidence. However, it isn't happening that way.

The quote stated earlier, "from what is to what may be" could refer to a metaphysical side to it. A much "broader view of nature," of complimentary views like this deist evolution theory with the Religion of IT, might provide clarity, "each serving different purposes and spurring complimentary insights."

On the Origin of Time, (Thomas Hertog, page 249);

> "I reflected on our journey. In our search for the ultimate underpinning of reality we had, in some kind of curious loop of interconnections, been led back to our own observations. "We are a way for the universe to know itself," Carl Sagan famously said.
>
> It seems to me that in a quantum universe — our universe — we are getting to know ourselves. Top-down cosmology, whether in its holographic form or not, is grounded in our relationship with the universe. There is a subtly human aspect to it. And I have had the strong feeling from a God's-eye to a worm's-eye view of the cosmos has felt like a homecoming to Stephen Hawking indeed."

Charles Ives, "The fabric of existence weaves itself whole."

30

The Cambrian Explosion

Rare Earth: Why Complex Life is Uncommon in the Universe, Chapter 7, pg.125:

> "Our planet was without animal life for the first 3.5 billion years of its existence and was without animals large enough to leave a visible fossil record for nearly 4 billion years. But when, 550 million years ago, sizable, and diverse animal life finally burst into the oceans, it did so with a figurative bang — in a relatively sudden event known as the Cambrian Explosion. Over a relatively short interval of time, all of the animal phyla (the categories of animal life characterized by unique body plans, such as arthropods, mollusks, and chordates) either evolved or first appear in the fossil record..." (pg.129) "... the discovery of the seemingly instantaneous appearance of complex fossils convinced most scientists…" (when it was discovered) "that life was spontaneously created — put on earth through the action of some deity; this observation is still cited by creationists as evidence against the theory of evolution…"
>
> [*However*, on the next page...] "Paleontologists have since proven Darwin correct, for the supposedly

> "barren" strata beneath those that bear what were thought to be the first fossil skeletons, do indeed contain the ancestors Darwin sought and theorized about. They were long overlooked or missed, however, because of their rarity or ridiculously small size. Most organisms from the youngest "Precambrian" time both were tiny and lacked skeletons, so they rarely left obvious traces in the fossil record. They are extremely hard to detect unless special processing techniques are used to extract them from their entombing matrix..."

This should end any further argument against the common ancestry of all life.

> (pg.131) "The Cambrian Explosion remains a relatively sudden and single outburst of animals — an unleashing of abundant and voracious creatures upon the earlier bacterial world, which continues, unabated, more than half a billion years later. With the exception of life's first formation, it remains the most profound biological event to have occurred on this planet."

> *Bart Ehrman*, (paraphrased); "For any given event without an explanation, an unlimited number of not provable explanations could be offered. None of them would necessarily be true but they can't be proven to be true or untrue. One person might ascribe a miracle to God while someone else might claim that extraterrestrials are responsible. Without evidence to back their claims, their explanations and opinions are meaningless. Many events that were initially thought

> miraculous, later turn out to have a reasonable explanation or fall into the category of unexplained phenomenon. For thousands of years, humans explained away anything that they didn't understand as being caused by a god or the God."

See what I mean?

There are creatures on Earth who have been unchanged for millions of years because they didn't need to adapt. Some species have adapted in separate ways, making them different from each other, but others simply haven't needed to change. Horseshoe crabs have been around for 300 million years with barely any change at all. The Horseshoe Crab hasn't needed to change. That phenomenon is reserved for species that have been through extinction events that directly effected their continued existence.

The evidence for random chance mutations is deceptively faulty. The evidence for intelligent design is deceptive and fabricated to enhance and enrich the lives of religious zealots extorting wealth off of the ignorance of the masses, using the fantasy of going to a heaven or hell. This is further perpetuated by atheistic science's insistence to always being right about everything, a form of hubris, arrogance, and delusions of grandeur; allowing religion to deceive the masses unabated throughout the ages because they make atheism look evil. This has occurred simply because way too few people actually question what has been indoctrinated and/or assumed by both sides.

How much further evolved we become depends on whether our species can survive our ontological dilemma between the selfish nature and a benevolent one. That is the test. Are we focused on humanitarian goals or the beachfront home? The beach? Oh, too bad! You lose the test, like Adam and Eve taking a bite off that poisonous fruit. In dying, you shall die! (explained in detail, Part 2)

We have figured out a lot over the course of human history and it's all happening because that MIND behind the designed design knew that it would happen. It is all designed to proceed as it has without guidance or interventions or random chaotic occurrences. You want proof? Look around you. Could we have created all this from all the materials at hand if not for having a perfectly designed planet and universe to be suited for accomplishing it?

We evolved into intelligence and sentience by questioning everything. In a perpetual judgment of our beliefs and practices, we strive to be better than our savage and ignorant history reveals. If humanity is going to survive their delusions, imaginations, and destructive practices they must question everything. No other living thing has the level of self-awareness that can question existence by independent intelligent thought processes, only us. Religion must strive to deny you that right and opportunity just to profit from your ignorance and maintain their political power and status as a profitable "industry." Yes, religion is a *business* and political movement that feeds off capitalism through exploiting and manipulating the ignorant. It's time to wake up and look at the reality of everything. What is desperately missing from humanity? *Benevolence*, plain and simple.

Rare Earth: Why Complex Life is Uncommon in the Universe By Peter D. Ward and Donald Brownlee:

> Pg. 36-37, "Earth is the only location in the universe that is known to have life, but it is only one of perhaps millions of habitats in our galaxy, and trillions in the Universe, which might harbor (simple) life. From the biased viewpoint of Earthlings, however, it does appear that Earth is quite a charmed planet. It has the right properties for the only type of life we know, it formed in the right place in the solar system, and it underwent a most remarkable and unusual set of evolutionary processes. Several of its neighbors in the solar system even played highly fortuitous, supporting roles in making Earth a congenial habitat for life. The near-ideal nature of Earth as a cradle of life can be seen in its prehistory, its origin, its chemical composition, and its early evolution. What are the most crucial factors that allowed Earth to support advanced life? Earth has offered (1) at least trace amounts of carbon and other important life-forming elements, (2) water on or near the surface, (3) an appropriate atmosphere, (4) a very long period of stability during which the mean surface temperature has allowed liquid water to exist on its surface, and (5) a rich abundance of heavy elements in its core and sprinkled throughout its crust and mantle regions."

> "Earth is actually the final product of an elaborate sequence of events that occurred over a time span of some 15 billion years, three times the age of Earth itself. Some of these events have predictable outcomes, whereas others are more chaotic, with the final outcome controlled by chance. The evolutionary path that led to life included element formation in the Big Bang and in stars, explosions of stars, formation of interstellar clouds, formation of the solar system, assembly of Earth, and the complex evolution of the planet's interior, surface, oceans, and atmosphere. If some god-like being could be given the opportunity to plan a sequence of events with the express goal of duplicating our "Garden of Eden", that power would face a formidable task. With the best intentions, but limited by natural laws and materials, it is unlikely that Earth could be truly replicated. Too many processes in its formation involved sheer luck... Just as the more familiar evolution of animal life involved many evolutionary pathways with complex and seemingly random branch points, the physical events that led to the formation and evolution of the physical Earth also required an intricate set of nearly irreproducible circumstances... Any construction project requires that building material be on site before the actual construction begins."

If a God was planning to make a habitat for breeding humans into a higher intelligence, the means to do it would need to be present before the breeding and gardening takes place.

31

Alone in the Universe

In his book *"Alone in the Universe"*, John Gribbin wrote,

> "change over time has accumulated in the same amount of time in all these species. This shows that the molecular clock has been ticking at the same rate in all species, ever since the split (from earlier forms)... So, we know that the clock has been ticking at the same rate in humans and the other apes."

"*Astrobiology: Exploring Life in the Universe*" by Lewis Dartnell,

> "Some researchers believe that the emergence of life is effectively inevitable, given appropriate conditions."
> "Surely the greatest enduring mystery of biology is the origin of life on Earth; understanding how you get from ubiquitous simple organic substances to replicating polymers and metabolisms. We really don't have much of a clue about the how, where, when, and how long. Defining the range of conditions that would permit life on other worlds is even harder… it is possible to estimate the likelihood of life arising, but Earth could be the only inhabited world in our neighborhood. However, considering the mind-boggling size of the visible Universe, it is tempting to

> argue that life must exist elsewhere, just perhaps beyond our reach. The chances are the whole Universe is teeming with life, we just need to start reaching out to find it."

The consequences of the natural development of organic chemistry is the function of the universe working naturally, internally, as it was designed to do rather than order emerges spontaneously from chaotic systems as Darwinism's atheist claims insist.

It might simply be that at its core, in emerging and evolving life, it is the result of the "natural development of organic chemistry," in an interconnected, living, super-organism like the universe. There's a natural function behind the design to emerge and evolve life. The emergence of life was inevitable given right conditions. So, if the emergence of life is the natural development of organic chemistry, none of that sounds like divine intervention or random occurrence theory.

The passages by Lewis Dartnell in his book says that some "supporters" of the research on astrobiology are suggesting that as Darwinism is concerned, "order emerges spontaneously from chaotic systems and patterns" but are actually a consequence of "the natural development of organic chemistry." in other words, Darwinism's cumulative selection or order from chaos is not how it happened.

A "hypothesis" is defined by Webster's dictionary as "an assumption made esp. in order to test its logical or empirical consequences." If we are just the "clay" of the potter, how can we know what the potter is trying to mold until the potter has molded us into the final product, the pot? We would not know, and we don't.

When I realized that my theory would be classified as "Deism," I read up on it.

(*Bing*) "The term is used chiefly of an intellectual movement of the 17th and 18th centuries that accepted the existence of a creator on the basis of reason but rejected belief in a supernatural deity who interacts with humankind."

(*Collins Dictionary*) Deism, "belief in the existence of a God on the evidence of reason and nature only, with rejection of supernatural revelation (distinguished from theism) 2. belief in a God who created the world but has since remained indifferent to it."

(*Rollins.edu*) Deism, "In the eighteenth century, known as the Age of Enlightenment or Age of Reason, deism emerged as a theological position that attempted to define the natural world and man's relationship to God through the application of Enlightenment ideals such as reason, rationality, and order."

It's not that a Designer of some kind is indifferent to us as individuals, sports teams, and nations for being the reason that IT is invisible, silent, and unreal to us. This apparent indifference seems necessary in evolving us into benevolent beings. For it to work, we must be unaware of why we exist until we evolve into a stage in which we are ready for the process to move forward. We must be free moral agents and individuals unaware of why we exist, for now. We had to evolve without knowing why. Without direct guidance.

Without intervention. IT couldn't reveal IT's existence to us before we're ready. It would not suit the purpose at all and would just throw a wrench at the wheels of the design. Why? Because we are clearly not ready yet. We find ourselves on the verge of burning up or blowing up our planet and act like it either isn't important — because we go to a heaven or hell anyway and/or their God is going to destroy everything and create "a new heaven and a new earth" — or that there's nothing that we can do about it because billionaires' control everything and don't care about future generations.

REVELATION 21:1-2, (NIV) "Then I saw a new heaven and a new earth, for the first heaven and the first earth had passed away, and there was no longer any sea. I saw the Holy City, the new Jerusalem coming down out of heaven from God, prepared as a bride beautifully dressed for her husband."

We can't create a peaceful cohabitation as long as we are killing each other over stupid shit. Because that is exactly what we are doing.

An Impersonal Universal Consciousness would need us to be benevolent, kind, compassionate and intellectually advanced much more than we seem to be now, or we'll eventually destroy ourselves. All that we can do at this point is just barely survive our instinct of survival of the fittest which has morphed now into a survival of the richest and finally Come Together, before it's too late. Because what a real "God" would expect and require from us to take part in the purpose behind the design is simple, ALL YOU NEED IS LOVE! That should be easy enough for most people to understand, but sadly, it may be impossible for many. *Matthew* 22:14, "For many are called, but few are chosen." Knowing what we can see with our own eyes about human nature, that statement seems correct.

The real Designer of the universe is not the God of the Old Testament! Here are a few examples of why and there are plenty of others. Keep in mind that the scribes who wrote this stuff wanted people to fear God and painted a picture of an authoritarian inhuman tyrant — which is not the image and likeness of a truly advanced and intelligent personage who might just be trying to raise a family through us.

THE FOLLOWING IS NOT THE REAL DESIGNER OF UNIVERSE. This is an evil monster that was invented by the minds of evil men…

DEUTERONOMY 13:6-10, (NIV) "If your very own brother, or your son or daughter, or the wife you love, or your closest friend secretly entices you, saying, "Let us go and worship other gods" (gods that neither you nor your ancestors have known, gods of the peoples around you, whether near or far, from one end of the land to the other), do not yield to them or listen to them, Show them no pity. Do not spare them or shield them. You must certainly put them to death. Your hand must be the first in putting them to death, and then the hands of all people. Stone them to death., because they tried to turn you away from the Lord your God, who brought you out of Egypt, out of the land of slavery."

NUMBERS 31:17-18, (NIV) "Now kill all the boys. And kill every woman who has slept with a man but save for yourself every girl who has never slept with a man."

1 SAMUEL 15:3, (NIV) "Now go, attack the Amalekites and totally destroy all that belongs to them; put to death men and women, children and infants, cattle and sheep, camels and donkeys."

That God is a fable, a farce, a fairy tale, and a reflection of man's evil tendencies being projected on to their primitive perception of a god. How can Christians knowingly worship this monster? How could they not know about the many verses from their Christian Bible that paints a picture of a horrendous, insane, and despicable God? Don't they KNOW what is in the Holy Bible? I'm sorry to tell you that most do not. They feel no need to know because they accepted what they were taught and told and must keep to it that way to their graves without question in order to prevent a depressing condition that deep down is gnawing at them in the mental anguish of cognitive dissonance. They know that their religious beliefs are messed up but their whole lives were devoted to their doctrine of beliefs.

32

Voltaire, the Deist Humanist

"*Those who can convince you to believe in absurdities can convince you to commit atrocities.*"

Yes, they have! Voltaire had it right:

> "Francois-Marie Arouet was a French Enlightenment writer, philosopher, and historian. Known by his nom de plume M. de Voltaire, he was famous for his wit, in addition to his criticism of Christianity — especially of the Roman Catholic Church — and of slavery."

(per *Wikipedia*) "Voltaire was a part of the Deist movement which is a belief in which God has created the universe but remains apart from it, permitting his creation to administer and steward itself and all creation through natural laws."

(*Google*) "(born November 21, 1694, Paris, France — died May 30, 1778, Paris), one of the greatest of all French writers."

(*Britannica*) "What was Voltaire's main philosophy? Voltaire believed above all in the efficacy of reason. He believed social progress could be achieved through reason and that no authority — religious, political, or otherwise — should be immune to challenge by reason."

A leader among intellectuals during the "French Enlightenment" at the same time that America was fighting for independence, Voltaire was a deist and probably the first well known and outspoken deist. He was dealing with a theist world in which you weren't allowed to challenge the status quo with facts, common sense or reason. Very much like the way most religions still operate.

Voltaire emphasized in his work the importance of tolerance, especially religious tolerance. He advocated freedom from and for religion, also for freedom of expression and was a pioneer in the cause of separation of church and State. He died in 1778, just two years after the signing of the Declaration of Independence in America, when my country's forefathers began the experiment in self-governing with an elected democratic republic, which we know is based on Voltaire's Deist and Libertarian (real liberty) philosophy (*not* the far-right kind of anti-government Libertarian).

I don't know for sure, but I'll bet many of those founding fathers of America were very familiar with Voltaire's work. It can be reasoned by the records of what they wrote that many were deist like Voltaire and would have read his writings. The founding fathers were mostly young men and Voltaire was old at the time of American independence. However, many of our greatest heroes from that period were very studious and learned and would be remarkably familiar with one of Europe's most acclaimed intellectuals. A declaration of independence that includes a clause to prohibit the government from enacting laws that recognize any religion was a view widely known about Voltaire.

It would only be important to agnostics, atheists, and deists, as well as believers of the least popular religions. An example of another reason it is so disappointing to see the United States voting for a party who has been taken over by Christian extremists like Republicans have. Where are the agnostics, atheists, deists, and believers in other religions when elections decide how they are governed?

I see the main cause of mass extinction as the obsessions and distractions of games, sports, entertainment, and dangerously aggressive antisocial politics. Like the Emperor who deals with a growing resentment of his dictatorship by holding gladiator games and blood sports to appease and distract the indoctrinated sheep.

(*History.com*) "Followers of Buddhism don't acknowledge a supreme god or deity. They instead focus on achieving enlightenment — a state of inner peace and wisdom. When followers reach this spiritual echelon, they're said to have experienced nirvana. The religion's founder, Buddha, is considered an extraordinary being, but not a god. The word Buddha means "enlightened."

Wikipedia definition of '*missionary*': "A missionary is a member of a religious group which is sent into an area in order to promote its faith or provide services to people, such as education, literacy, social justice, health care, and economic development. "

That is the more modern and Christian description of Missionary because the original Jewish Jesus movement and Buddhism were not looking to convert anyone to a "Faith."

Jesus and Buddha were living examples of how to live in tune with the universe that made us. And as we know from the Gospels, Jesus taught his followers to provide services to people in the most need with compassionate, merciful social justice. The hanging judge must be replaced with a benevolent one if humanity is going to continue to evolve.

History's greatest humanitarians — historical Jesus as well as the embellishments in the Gospels being the best example — were in favor of a socialist type of economic system and against the idea of allowing a few to be enormously wealthy just to horde and gorge themselves without a care for the suffering it causes. We can see the evil of that in contrast to socialism. It pains me to no end to see a lifetime of greed controlling the world when that was not meant to be if Christianity had followed the Jesus of the Gospels instead of the licentious lunatic Apostle Saul/Paul and the many of his followers who wrote the New Testament Gospels and Epistles.

The Impersonal Universal Consciousness would want all humans to be healthy, happy, and safe (the historical Jesus) because IT wants us to continue to evolve into benevolence. We are not evolving right now; we are devolving.

> "What is faith? Is it to believe that which is evident? No. It is perfectly evident to my mind that there exists a necessary, eternal, supreme, and intelligent being. This is no matter of faith, but of reason." ~*Voltaire*

33

Survival of the Richest

To most people, the term "evolution" brings to mind the discovery in recent centuries of human origins. Whether our species branched off of the same ancestors as apes, chimps, and other "primate" creatures, the facts are self-evident to any reasonable person that we live and act very much like those animals because we were animals that evolved from the animal kingdom into this human kingdom. Other than the obvious physical similarities and differences, what is the most striking difference between us and all other animals? Simple observation should tell you that as animals cannot control their instincts, we can!

Merriam/Webster Dictionary's definition of "instinct";

1. A natural attitude,
2. A largely inheritable and unalterable tendency of an organism to make a complex and specific response to environmental stimuli without involving reason.

This "natural attitude" of animals and humans are basically the same. The difference is our ability to alter our responses to it with or without "reason." It is our free moral agency to decide on our own actions. Has humanity been fully using our ability to "reason" in dealing with others and the planet that we live on? Not much! Is it wise to ignore our ability to "reason" in favor of using our animal instinct to fight and kill each other like wild animals? Religion has the word "righteousness" confused with being blindly obedient to their perception of a god. The word should instead mean, "right-wise-ness." The condition to see things with the true wisdom of being correct about what is truly right, not blind obedience to some fictitious god.

Animals compete against each other for resources while predators kill other animals for food. Carnivores can't feel empathy toward their food. We are omnivores who can feel empathy, but often choose not to for selfish reasons. The real meaning of life really does matter if the survival of our species is now at risk. We know that it is. Are we just going to blindly burn up this planet and cease to exist as a species without asking the deepest and hardest questions imaginable because of the foolish system of faith? And is there something more to life than anything we have been told, sold, taught, sought, and fought over?

Why did it have to come down to some Armageddon, an event that was mistranslated in *Matthew* as, "end of the world" but really says "end of the age?" Moreover, some kind of apocalyptic, Earth-shattering scenario may be fast approaching in spite of us knowing that it is coming and with many people looking forward to it. It seems to have been unavoidable. It's as if it was humanity's destiny to cause a mass extinction event that threatens their very existence.

We were designed to use carbon burning fuels to power the machines that we were destined to make. We were destined to build this civilization and then procreate billions of humans with selfish billionaires running everything. We were designed to tap into the power of the atom. We were designed with apathetic natural selection, "survival of the fittest," which we use now to create a society of survival of the richest. We are designed and destined to run out of things, poison almost everything, and damage our climate, oceans, and mountains in the process to obliviously "be fruitful and multiply" ourselves to death. I know why Christians ignore our planet's peril.

Most are expecting and counting on a second coming and Day of the Lord or "Day of the LORD", or a rapture or the belief that everyone is going to a heaven or hell when they die. Practically all religions have had stories of a divine being or divine messenger prophesied to save them from whatever is threatening them. The story of Moses is made up. It never happened then, and we cannot logically expect it now, no matter how dire our situation is. It's ironic that now they are hoping for a Messiah to save us from destroying ourselves. They should know by now that it is we ourselves who are the threat to us now.

Unless one does the actual research into this subject at the level similar to or beyond what I have, their opinions and views are "subjective"; ("Merriam/Webster's Deluxe" paraphrased) "subjective" is peculiar to an individual; modified or affected by personal views, experiences, or backgrounds, and arising from conditions within the mind itself, indoctrination, and dogma — not attributed to the external stimuli of any actual critical study, extensive research, and unbiased review of the facts in historical contexts.

Our planet in its infancy went through 5 stages of heavy bombardment from the cosmic debris that created it. All those elements raining down like asteroids and meteors brought with them everything that would be needed to create a planet perfectly suited for life. It has been debated on how our oceans were formed. Consensus reveals that it was mostly from the cosmic debris that formed our planet to begin with, including meteors of ice. With the four top elements here being Hydrogen, Oxygen, Nitrogen and Carbon — the combination of Hydrogen and Oxygen forming water is another logical reason that we have the oceans, seas, underground aquifers, and springs. This may have included, "organic substances delivered from outer space accumulated and, perhaps, replication molecules emerged as cells took a tentative foothold."

(Lewis Dartnell) "*Astrobiology: Exploring Life in the Universe*" (2011) goes on to describe "It's startling just how quickly life seems to have sprung up."

(*Space.com*) "Lying at the very heart of the Milky Way is a supermassive black hole called Sagittarius A. About 4 million times the mass of the sun, this beast consumes anything that strays too close, gorging on an ample supply of stellar material enabling it to grow into a giant."

We are evolving in complexity toward a higher consciousness for all humanity.

If a MIND behind the design of everything has emerged and evolved us for a purpose, it would be illogical to allow practically anyone into a divine realm untested and still acting like the animals we once were. Michael Talbot's book comes from the mindset of the immortal soul doctrine. If we are evolving toward the development of a higher consciousness, or evolving into ET, or the realm of some kind of spirit world is pushing itself into matter and back to a spirit world again — no matter what you might believe — the truth has to be a logical, practical, empirical, and believable purpose that humanity is being led into that is impersonal, logical and practical. I'm not seeing that from any science or religion.

34

Significant Quotations

Carl Sagan, “You can’t convince a believer of anything; for their beliefs is not based on evidence, it’s based on a deep-seeded need to believe.”

Socrates, (possibly by him) "For I was conscious that I knew practically nothing."

Steven Weinberg, “In the ordinary moral universe, the good will do the best they can, and the worst will do the worst they can but if want to make good people do wicked things, you’ll need religion.”

Ricky Gervase, “Religion’s greatest trick wasn’t convincing people that there was a God who was all powerful, it was convincing everyone else that you couldn’t ridicule the idea.”

Astronaut *Edgar Mitchell* (After being in space) “You develop an instant global consciousness, a people orientation, an intense dissatisfaction with the state of the world, and a compulsion to do something about it. From out there on the moon, international politics look so petty. You want to grab a politician by the scruff of the neck and drag him a quarter of a million miles out and say, ‘Look at that, you son of a bitch.’”

Bart Ehrman;

> "The fact that the Gospels claim that Jesus called himself a divine being doesn't mean the historical Jesus himself did." {he didn't}
>
> ("*How Jesus became God: The Exaltation of a Jewish Preacher*") "Almost all the evidence shows that it was orthodox scribes who modified their texts in order to conform more closely with orthodox theological thought."
>
> "The idea of the Rapture has not been taken from the Bible; it has been read into the Bible."

Ethan Allen, "Those who invalidate reason, ought seriously to consider, Whether they argue against reason with or without reason; if with reason, then they establish the principle, that they are laboring to dethrone, but if they argue without reason, (which, in order to be consistent with themselves, they must do) they are out of the reach of rational conviction, nor do they deserve a rational argument."

35

Agnostic Secular Deist

Macmillan's Dictionary, "AGNOSTIC, a person who believes that nothing is known or can be known about the existence of God."

Webster's Dictionary,

> "SECULARIST, an upholder of secularism; one who theoretically rejects every form of religious faith."
>
> "DEIST, one who acknowledges the existence of a God, but denies revealed religion; a freethinker."

A simple definition of agnostic would also be, one who is not committed to believing in either the existence or nonexistence of the God of religion or any kind of god who involves himself with human affairs. That much I can be sure of. It is a question that we cannot fully answer at this time. Therefore, we cannot believe anything on the subject when no concrete empirical evidence has been presented. However, I agree with Thomas Edison, that while I don't believe in any god/God of religion, I have no doubt that there is an intelligence involved behind in the existence of the universe.

When certain people believe that they have been permitted to "rule over the fish in the sea and the birds of the air, over the livestock, over all the earth", (*Genesis* 1:28 NIV) by their

Creator, they don't respect those things that they've been given control over; taking for granted and then abusing at their leisure and pleasure. They believe that the universe was made just for them to use. This mindset, whether realized or not, is that God has given everything directly to them to use however they want. Besides which, space, time and matter are just temporary things compared with their imagined, fictitious immortal souls or their fervently imagined "free gift of eternal life" because of a human sacrifice to a god.

Now that I know what I know, which they don't know, I can't believe so many modern-day people today believe that scam of all scams. So, why bother caring about our environments? These animals, plants, fish, trees, and future generations are not important at all to the most religious Christians among them. Some believe that there will be plenty of it for the future generations, particularly after our "heavens and earth" are passed away and a "new heavens and a new earth" comes down (*Revelation* 21:1-2) from the sky with a "New Jerusalem." That was/is the main excuse for wasteful destruction of resources. They believe that they'll all go to a heaven or hell or witness the imminent (it's *always* imminent) and expected, prophetic, apocalyptic, and fictitious "end of the world."

Like the statements made in the Gospels allegedly by Jesus, and throughout the history of civilization, there have been reforms of cultural standards toward empathy by those who believed that a God was directing them in caring for the needy and service to others. It seems clear that humanity has been increasing in empathetic benevolence because of history's leaders of human rights and philanthropy, until lately. Jesus would be a good example of that.

But what do I know? Socrates said something that opened up my mind to how I should approach the existential ontological subject.

What is Socratic ignorance? Socratic ignorance is a kind of awareness of what one does *not* know. It is captured by the well-known statement:

"I know only one thing — that I know nothing."

Socratic ignorance is also referred to as "*Socratic wisdom*". It is true wisdom at the core of intelligence in a firm realization that we don't know anything at all until we learn it somehow and therefore are basically ignorant of the deepest meaning in life, love, liberty, and especially the secrets of the universe.

36

Geoffrey Riggs

This next chapter was offered to me by an individual I know from social media. This is his take on the humanitarians of history.

> "The reformer ultimately generates some groundbreaking principle addressing some part of the socially marginalized in a new and more inclusive way. It's from counter-cultural crucibles like this that most of the more inclusive paradigms for the marginalized have emerged for thousands of years.
>
> I still don't credit any creed, doctrine, or religion. But it's now very evident that the most powerful urges to help our fellow creatures in pioneering and path-breaking ways mostly come from figures who feel — or imagine they feel — some sudden visceral closeness to a beneficent presence duly sensed as divine.
>
> But whether or not such personal, visceral, and counter-cultural beliefs of the divine are delusional or based entirely on a true reality that only the most altruistic and empathetic can see, it is plain that humanity requires such "delusions" constantly in order to function with any modicum of continually expanding social justice."

Truer words are rarely spoken. If this is the case, the increase of empathy in humanity over the millennia is attributed to social reformers inspired by what they perceived as their mission from God; in spite of the authoritarian, apathetic, and delusional control religion demands on everyone, by setting up and controlling the narrative.

Humanitarianism might very well be a function of a deist designed evolution toward the development of a benevolent nature growing in humanity in preparing us for whatever reason the Designer has for emerging and evolving life. There is no logical evidence in a guiding and directing God persona pushing our buttons and turning the dials of the universe. That God doesn't exist, but some kind of MIND did design and implement this living universe for a purpose that is beginning to awaken those who can sense it.

No government should be guided by the charlatans of Evangelical Christianity, or Zionist Judaism, or the Islamic Republic. Even the military government of Myanmar has consistently violated the peaceful precepts of Buddhism. A totally secular world would be the only peaceful one. One that sees all of us on this one planet as family. The only way to a peaceful world will require a thoroughly secular one and humanist.

Nothing else will work.

Arthur Conan Doyle in *Sherlock Holmes*: "When you have eliminated all which is possible then whatever remains, however improbable, must be the truth."

Fred Naiden, History Professor at UNC/Chapel Hill, among his many books is this passage...

> "Religion dominated warfare because gods dominated everything," — "All ancient commanders played religious parts, but Alexander [the Great] played the most. He uttered or inspired the most prayers and made the most sacrifices, and he did so in the most places, languages, and rituals. We may think of him as the Pope, the Holy Roman Emperor, and the head of the Church of England, all in one."

It was during the Greek occupation of Judea well before the Romans took over that Jewish doctrine embraced the concepts of a devil and demons as well as other influences from Greek mythology.

The very concept of war is completely against and diametrically opposed to the teaching of the historical Jesus from the Gospels and all other notable humanitarians in human history.

Matthew 5:9, (NRSV), Blessed are the peacemakers, for they will be called children of God."

Where are they now? Has there ever been any real peacemakers other than the victors of war? When peace is achieved at the end of a sword, gun, or bomb, it is never going to be truly peaceful. There will always be those on the losing side who feel cheated and treated like less than human by the victors.

"*Stephen Hawking's Final Theory*" is two-fold; partially involving the Holographic universe theory and something referred to as "top-down cosmology." One hundred years ago, the fellow physicist and colleague of Albert Einstein, David Bohm, held long conversations about holographic universe theory with Einstein. Einstein said that Bohm had explained the principles of the theory better than anyone but in spite of that, he was still not convinced that the theory was true.

PART II

The Religion of IT

37

KILL! KILL! KILL! Really?

Kill the witches (*Exodus, 22:18*)

Kill adulterers (*Leviticus, 20:10*)

Kill blasphemers (*Leviticus, 24:14*)

Kill false prophets (*Zechariah, 13:3*)

Kill fortune-tellers (*Leviticus,* 20:27)

Kill non-Hebrews (*Deuteronomy, 20:6-17*)

Kill sons of sinners (*Isaiah, 14:22*)

Kill gays (*Leviticus, 20:13*)

Kill non-believers (*2 Chronicles, 15:12-13*)

Kill anyone who curses God (*Leviticus, 24:16*)

Kill all males after winning a battle (*Deuteronomy, 20:13*)

Kill disobedient children (*Exodus, 12:17 & Mark, 7:10*)

Kill strangers close to a church (*Numbers, 1:48-51*)

Kill those who work on the Sabbath (*Exodus,31-15*)

Kill any bride discovered not a virgin (*Deuteronomy, 22:21*)

Kill those who curse their father or mother (*Leviticus, 20:13*)

Kill those who worship the wrong god (*Numbers 25:1-9*)

Kill those who kill anyone (*Leviticus, 24:15*) [HA!]

REALLY? Kill, kill, kill? *Is that your God*?

Religion feeds dangerous delusions! This means that people depict gods as mostly inhuman, heartless, barbaric, cruel, spiteful, unmerciful, ruthless, malevolent, tyrannical monsters. That might be the god/God that you thought is the intelligent designer of this whole universe because you were indoctrinated and brainwashed! Really? It's such a huge mistake that people in, "The Year of Our Lord, 2025," worship that nightmare of a "LORD God" from the Old Testament. That LORD God thing is not what a real MIND behind the design of the universe would be, whether guiding and directing things or inactive and just waiting for us to evolve more. What would a real MIND behind the design of this universe naturally be? A real God would not be composed of this material substances that we are. It would not have that same natural selection, survival of the fittest, self-centered attitude of human beings like we have. All the gods envisioned by humanity have been depictions of human-like characters.

A real God, if there is one, would be a benevolent, empathetic, merciful, and logical MIND of ultimate intelligence and right-wise-ness {aka, "righteousness" and "right judgment," true "wisdom"} — but purposely enigmatic and elusive by design to keep us in the dark as we evolve, an Impersonal Universal Consciousness. The one human being in history who best exemplifies what that "MIND" behind the design would be is the Jesus of the Gospels, minus the forged and fictitious evangelical and political additions in the text and narrative.

We all must realize by now that the universe seems to be a grand invention of magnificent design by some kind of MIND WHO is not involved at all with us and never has been. The concept of evil simply comes down to being caused by the absence of good, just as darkness is the absence of light. Unrestrained greed and selfishness is the beginning of what is evil in the eyes of a MIND behind the design. The Jesus of the Gospels exemplified the opposite of selfishness. Selfishness is ultimately destructive and unmercifully deadly. It is the essence of, the cause of, and the basis of evil incarnate. Here is some thoughts on the subject by Geoffrey Riggs.

Geoffrey Riggs, "Ultimately, the question of where evil comes from is unavoidable. Is evil merely the absence of good? Is it a faulty prefrontal cortex that is not functioning at full capacity, allowing the less evolved reptilian part of the brain to take over? Or is evil a proactive force that can impose its will on others if given half a chance? Is it something external that is always seeking recruits?

Candidly, in addressing this question, I happen to place a lot of stock in modern textual research. Some may be made uncomfortable by it; I welcome it. An interesting window on these matters, like personifications of evil such as Satan, etc., opens up when we weigh some of the most recent textual findings in the Synoptic Gospels. A strong indication of the most historical strata in the Synoptics is the presence of what are termed Aramaic stylometric in the original Kione Greek texts. These are traces of obstinate Aramaic grammatical and etymological structures fitted all too awkwardly in the original Greek versions of specific verses in the Gospels. What their presence indicates is an extremely early oral source for such verses. In other words, if there are any verses that go all the way back to eyewitness material, it is highly likely these verses are the most authentic. Drilling down further, the verses most imbued with Aramaic stylometrics are apparently that distinctive stratum of parallel Jesus sayings found in tandem in Matthew and Luke. These parallel sayings all relate to each other in inextricable ways and are mostly very down-to-earth, blessed be the poor, you cannot serve God and mammon, the last shall be first, etc. A highly distinctive vocabulary with a whole array of characteristic turns of phrase and linguistic habits distinguishes them from the surrounding material.

If any stratum reflects Jesus's own voice directly, it's very likely these parallel verses. What is especially surprising is that amid these down-to-earth verses is one atypical passage in particular with all the same vocabulary and linguistic signposts as the parallel down-to-earth sayings in Matt./Luke: the passage describing Jesus's early encounter with Satan in the wilderness! Like the parallel sayings in Matt./Luke, this passage is also present in both these Synoptics, and found in both these two almost word for word. Like those other passages, this encounter also has some traces of Aramaic stylometrics.

But it is practically the only such parallel passage with elements of the supernatural. Yet it emerges as imbued with exactly the same "voice" as the down-to-earth parallel sayings that form 99% of this distinctive Aramaic-tinged material. What this suggests is that this encounter comes from Jesus's own mouth. Is this the most authentic example of a first-hand account of Satan as something all too real? Or if it is an authentic first-hand account, might it instead reflect the impressions of a susceptible human being whose faculties, depleted by 40 days of virtual starvation, were playing tricks on him?

Yet, was Jesus susceptible to such confusion? Was this instead what he was seeing for real, and is this the most concrete and authentic example we have of the way Satan toys with us? He certainly plays ruthlessly with Jesus's ego here. Is that the key to evil? If you have a fragile ego, are you more susceptible to evil — and/or Satan? Are the fragile egos of those hypnotized by demagogues the key to the way democracies can collapse? These are some of the troubling questions that arise for us today.

38

Ride the Waves of Knowledge

Britannica: {There were many different kinds of "Christologies" during the first centuries depending on your previous religion}

> Definition of 'Christology': "Christian reflection, teaching, and doctrine concerning Jesus of Nazareth. Christology is the part of theology that is concerned with the nature and work of Jesus, including such matters as the Incarnation, the Resurrection, and his human and divine natures and their relationship."

No, sorry Britannica, that is technically not a good definition, it doesn't describe what the word means to scholars.

A "Christology" is what scholars call the different groups of very early followers and believers of the first Christologies (the first one was Jewish) of those first few centuries before the Roman Empire adopted the most popular one in the 4th century. The many versions of early Christianity had very different opinions of Jesus before that 4th century adaptation by the Roman Emperor Constantine who made a distinct and particular version of it into the official religion (the Orthodoxy) of the Roman Empire.

I am reminded that knowledge for us comes about in waves. When one generation ignores new facts revealed (for whatever reasons), the following generations discover the same facts and eventually, it catches on with the general population. Religious people have a long history of ignoring facts. It's the truth. However, science does not have the answers either because they reject certain facts that doesn't fall in line with their atheist doctrines. The mysteries of quantum physics and holographic cosmology will not be solved until scientists realize that it came from a highly advanced MIND from somewhere beyond the limits of our 3-4 dimensional universe. Quantum Physics and Metaphysics will converge when both sides can no longer ignore the potential benefits of evolving further complexity in our species through the designed design of the interconnected universe by a MIND of some kind for exactly that purpose.

The Gospel writers wanted it to look like Jesus was drawing people to him in order to join his religious cult and they embellished their third-hand narrative stories to reflect that agenda. But that isn't the case. At that time, Jewish sects that were Messianic in nature were exclusionary (not seeking converts) while anxiously awaiting someone to come and lead their army against Rome to take over Judea or witness some miraculous messianic apocalyptic event to destroy their Roman overlords.

From what I can tell now, from the oldest sayings attributed to Jesus, he may not have been trying to enlist people to his Messianic cause but instead was trying to teach about what we must do ourselves individually to enter the kingdom of

God/heaven and obtain a life after death of some kind. He just came from being baptized by John the Baptist and spending time alone in the desert to proclaim that the kingdom of God was coming soon and wanted to instruct us on how to prepare ourselves for that.

But Orthodox Christianity as of the fourth century wanted it to be about a salvation from sin by the sacrificing of a divinely born Jesus to his alleged father, the god of ancient Judaism. This was motivated by Apostle Paul's teachings of not wanting to adhere to Jewish laws and statutes by Christians and Jews alike. The original Jesus sect was an apocalyptic messianic and fundamentalist (to some extent) Jewish sect that the Jesus movement belonged to all along. Then after the crucifixion, under the leadership of Jesus' brother James, it was based on what Jesus and his Jewish sect taught. It included obedience to their God as a secondary requirement after adopting the altruistic nature that Jesus stressed in his sayings and deeds — having a deeply caring concern for others even to the detriment of oneself. That would have been his philosophy and ideology for what his god would require form us.

It was common practice for ancient religions to kill a human or animal in a blood-letting death ritual in order to satisfy a god. The idea is that when things go bad it is because a god is angry and punishes us for doing something wrong. The way to please and appease the god is a sacrifice. It seems dumb to us now, of course, but it made sense to primitively ignorant people. It's an example of how inhuman and ignorant humanity really was. Why does modern humans with a wide collection of knowledge still hold on to something as primitive as believing in gods and especially believing in a human sacrifice, a sacrificial lamb, like the violent death of Jesus? Think about that. Please.

> *Deuteronomy 14:22*, (NKJV) "You shall truly tithe all the increase of your grain that the field year by year." It is like a cut of your income that your God requires from you, ten percent, of whatever your income might be. That was the main motivation of Apostle Paul, in my opinion, to form his own Christology. There was also "offerings" that had to be paid, which usually means another ten percent on top of the tithe. It is also the main motivation of all Christian organizations. Definition of "tithe," Webster's Ninth New Collegiate Dictionary, "to pay or give a tenth part of esp. for the support of a church."

That's what Apostle Paul did by luring the Greco-Roman people to himself. Paul formed his Christology in an Evangelical style. The original Jesus movement had been exclusively Jewish under Jesus and his brother James but where forced into seclusion by the Jewish war against Rome only about 30 years after Jesus' death. Exclusive only to Jews and only if they were righteous and wise enough to repent (turn away from sin, of breaking God's laws) before that "Day of the Lord" comes that they were expecting was at hand. But then the Jewish war against Rome began about 30 years after the crucifixion of Jesus and any Jewish Messianic group would have been considered a threat to the Roman occupation. Roman armies harassed and detained any Jewish man connected to a messianic, apocalyptic sect. That was the point in which the original Jewish Christology dispersed and began to fade away. At that same time, you had Christologies of various kinds including Gnostics, "Marcionites", and Pauline followers seeking converts by writing third-hand narrative stories about Jesus with a great deal of embellishment. Much of it became predominantly embellished as the stories that were told and retold mixed with ancient pagan mythologies. In the coming decades into centuries, any writings of the original Jewish Christology were banned by the Orthodox and proto-Orthodox "Church." There are no preserved manuscripts by the early Jewish Christians. Only a few words here and there was written about them by orthodox historians, some of these church historians are called the church fathers.

“Judaizers” were what Paul’s phony Christology called one of the Jewish Christian sects that remained from the original Jesus movement:

> per *Wikipedia*: “The Judaizers were a faction of the Jewish Christians, both of Jewish and non-Jewish origins, who regarded the Levitical laws of the Old Testament as still binding on all Christians. They tried to enforce Jewish circumcision upon the Gentile converts to early Christianity and were strenuously opposed and criticized for their behavior by the Apostle Paul, which employed many of his epistles to refute their doctrinal errors.
>
> This conflict between Paul and his opponents may have been the reason for the Council of Jerusalem.”

Those statements are based on the book called ACTS that comes after the Gospels in the New Testament (NT). When there are competing political or religious ideologies through historical periods we see the history of it being written by the victors — the one that survived the conflicts. The Jewish War scattered the original Jesus followers as the competing and more popular licentious Christologies outlasted and outlived the original Jesus movement, the altruistic Jewish apocalyptic groups.

The real conflict between Paul and the Jerusalem church was not just about circumcision. It covered every aspect of Old Testament (OT) laws and statutes that Paul wanted to eliminate them all from being required by anyone. He makes that very clear from the writings in the NT that we know were from him. The result was a complete split between them. The Jewish Christians stayed true to the disciples and their original teachings of servitude to others and obedience to their God as Paul's self-indulgence and licentious doctrines became more popular with Gentiles from all over the Roman Empire. There is clear evidence in his writings that Paul developed a deep animosity toward all Jews, in spite of him being one of them. He got to the point where he believed himself to be above the law. You can see evidence of that in the chapter of this book called, "I Myself Am Not Under the Law."

"*Excavating Jesus*" by John Dominic Crossan and Johnathon L. Reed,

"Such a change from non-kosher for all to kosher for all, Paul twice condemns as hypocrisy. The assembly's response to Paul is not given but, no doubt, they would have said that no Paul, it is not hypocrisy but simply courtesy. Fourth, this is a pragmatic [definition, "practical, testing the validity of all concepts by their practical results"] question against which Paul mounts an in-this-case irrelevant argument. Had Peter and the others ever believed that kosher was still mandatory for their salvation, they could not so easily have omitted it...

Finally, to this pragmatic question Paul introduces "works of the law" three times and opposes to them "faith in Christ" three times. Paul's antithesis [a contrast or opposition, as of ideas, exact opposites] of faith and works might be theologically justifiable in the abstract but for that pragmatic question at Antioch it was irrelevant in the concrete. Does anyone believe that James, Peter, Barnabas, and all the others (save Paul) had opted for justification by "works of the law" rather than "faith in Christ"? Paul's position (at least as recorded to the Galatians) was akin to machine-gunning butterflies. James, Peter, Barnabas, and all the others who agreed with him {James}, were right at Antioch. Paul was wrong at Antioch."

GALATIANS 1:18-19, "Then after three years [after his "conversion"] I did go up to Jerusalem to visit Cephus [Peter] and stayed with him fifteen days, but I did not see any other apostle except James the Lord's brother."

Scholars know that he was not talking about a "brother" here in the same way as "brethren" or "believer" or "Apostle" can be translated as "brother." It was Jesus' real brother from the brothers of Jesus stated in Mark's Gospel (MARK 5:3) and also translated, "James, the brother of Jesus." The same brothers who that gospel says did not support Jesus. However, that was written by the people who were pushing for the antisemitic agenda in the Gospels in favor of the Pauline Christology of eliminating the teaching of keeping Gods laws.

The Brother of Jesus: and the Lost Teachings of Christianity, by Jeffery J. Bütz, pg.164-166 "Mainstream Anglican scholar Bruce Chilton, one of the organizers of the international Consultation on James, has come to the conclusions that James was indeed a Nazirite, that he most likely had at least some connection with this strict sect, and that this is the most likely reason that Jesus was called Jesus "of Nazareth"..."

"(Wikipedia) In the Hebrew Bible, a nazirite or a nazarite (Hebrew: נָזִיר Nāzīr) is a man or woman who voluntarily took a vow which is described in Numbers 6:1–21. This vow required the nazirite to: Abstain from wine and all other grape products, such as vinegar and grapes. Refrain from cutting the hair on his head." (Wikipedia) "These passages may indicate that Jesus intended to identify himself as a nazirite ["not drinking the fruit of vine"] before his crucifixion."

This would indicate evidence that the Last Supper never happened. Any honest examination of the Last Supper story leaves little doubt that it was an adaptation of a Gentile mythology's religious practice. It's an example of how Greco-Roman mythological religious rituals blended in with the Pauline Christology during the centuries leading up to the Orthodoxy under Catholicism. We know that the practice of drinking wine as a symbol of drinking the slain blood of a deity goes back well before the time of Jesus.

"*The Brother of Jesus: and the Lost Teachings of Christianity*" by Jeffery J. Bütz:

> "The answer is quite obvious, once one sees the larger picture that comes into view when all of the puzzle pieces are put together. But the emergent picture is not easy for many Christians to take in all at once. Now that we have completed the puzzle, we find ourselves facing a revolutionary (not to say heretical) paradigm— that not only were James and the apostles thoroughly Jewish in their beliefs and practices, but so was Jesus: the original orthodoxy was in fact a strict form of messianic Judaism…."
>
> "It is, above all, the Jewishness of Jesus to which modern research into James points; and, as we shall see, it is the Jewishness of Jesus and James that was a major reason for James's leading role in the early church being suppressed; and it is still the reason why the findings of recent research into James continue to be resisted by most Christian scholars."
>
> "But my overriding goal all along has been a quest for truth – no matter where it may lie, and no matter what inherited dogmas may need to be abandoned in order to attain it."

pg.10 of "*Misquoting Jesus: The Story of Who Changed the Bible and Why*" by Bart D. Ehrman:

> “James was considered by many early Christians to be the first “bishop” of the Church, the successor to Jesus following the crucifixion, making James in essence the first “pope”, not Peter as Catholic tradition has maintained. The church father Clement of Alexandria in his work Hypostases (Outlines), written at the beginning of the third century, makes the following rather startling statement: “After the ascension of the savior, Peter, James [the Son of Zebedee], and John did not claim pre-eminence because the savior had specifically honored them, but chose James the Just as Bishop of Jerusalem...”
>
> “James the Just” is a title for Jesus' brother commonly used in early church writings, “bestowed upon James by early Jewish-Christian groups such as the Nazoreans, the Ebionites, and the Elkesaites, who revered James for his outstanding righteousness under the Law and considered him to be the leader of the apostles after the death of Jesus. These Jewish-Christian sects claimed to be the remnants of the original Jerusalem church which had been scattered and dispersed after the siege of Jerusalem by the Romans in 66 CE…”

DID JESUS EXIST? *The Historical Argument for Jesus of Nazareth* by Bart D. Ehrman, “the earliest Christology of them all, probably the original followers of Jesus and so stems from the earliest Palestinian Christian communities… in the early 30s Common Era.”

Exactly after the same time that Jesus was crucified! Doesn't logic tell us that they would have known more than anyone who Jesus was or thought that he was and why? Of course, but religion has been hiding those facts. I shit you not! Why? Because Jesus was one of the leaders of a purely Jewish sect with Jewish beliefs, not what Christianity later became. Why? Two main reasons. Partly because of Paul's trickery and deceptive maneuver to create his own religious sect in the mold of a Greek Mythology and because of the Jewish war for independence from Roman rule. The Gospels were not written by anyone associated with Jesus while he was still alive. It was followers of the Pauline Christology decades later, of the doctrines that were established by a man who hunted down and persecuted the disciples a few years after the crucifixion, Saul of Tarsus, who became known as Apostle Paul to his followers, following the faking of a conversion event of claiming to have been spoken to by Jesus from the sky, in order to infiltrate the disciples and turn them away from being apocalyptic and messianic and to form his own sect that eliminated the requirement of keeping any of the laws and statutes of the Jewish God.

39

The Real Christmas Story

First, let's take a closer look at the origin of "Christmas." It takes place during the last week of the "Roman" calendar year just after the winter solstice and falls near the same dates as a pagan Roman festival of rituals called "Saturnalia" that has its origins in Roman and Greek mythological religions. The gullible masses of mythologies were led to believe that if they did not pander to the whims of a "Sun-god" and/or the god called "Saturn" during the coldest part of the year, there would be no spring.

But putting that aside for the moment, what can we discern from NT accounts of the "Virgin birth" of Jesus? Specifically, does the date of Christmas in the winter in December agree with the accounts of the birth of Jesus in Matthew and Luke's Gospels? No!

The Gospel called "Matthew" has a distinctly differing account of the birth story of Jesus from the account in the Gospel called "Luke." The other two Gospels, Mark and John, make no reference to Jesus' birth in any way. Those other two Gospels simply begin the story of Jesus from his baptism, which is exactly where the Jewish Adoptionist Christians would have also began their version of the story of the exaltation of Jesus from a man to God's firstborn son. It is interesting to note in the Gospel called "Luke" what time of year the virgin birth of Jesus is reported as.

> *LUKE 2:7-8*, (NKJV) "And she brought forth her firstborn Son, and wrapped Him in swaddling clothes, and laid Him in a manger, because there was no room for them in the Inn. Now there were in the same country shepherds living out in the fields by night."

There are several points to make here that indicate that it could not have occurred in the winter in Bethlehem. First, the term "swaddling clothes" was a reference to thin rags. It was written that way to convey poverty. But no parent would wrap a newborn in just that, if outside in the winter.

Second, even 2000 years ago, people who operate an "Inn" in Bethlehem would not turn away a very expectant mother. But if the Inn had "no room" for them it would most likely mean that the area was celebrating one of the 7 annual "Jewish" Holy Days at that time, which all fall between spring and autumn — none in the winter.

Thirdly, "shepherds" did not live out in "the fields by night" in the winter in Judea 2000 years ago or today.

Finally, a "manger" is a shabby old barn that would be very cold in Judea, barely a windbreak. Scholars can only guess that this birth story could only have taken place, if at all, between late spring to early fall.

Other parts of both Gospel accounts of the "virgin birth" story shows unambiguous evidence that the whole thing is a complete fabrication. Plus, Jewish sects of that era did not celebrate birthdays. It was more important to them to remember the date of someone's death.

In fact, the only place in the entire New Testament where Jesus is said to be divine from his birth or born of a virgin are those two accounts in the Gospels and that's all. There is no hint of his birth anywhere else either in the New Testament or in all historical records. In fact, there is only controversial evidence of Jesus existing at all.

So, when people say the meaning of Christmas has been distorted by capitalism as a boost to the economy during a time of the year when it's stagnant, there's a lot more about Christmas that is corrupted than just that. In fact, it was that time of year in late December that a festival in honor of the god of Saturn took place for a week and they called it "Saturnalia."

> (*WIKIPEDIA*) "Saturnalia is an ancient Roman festival and holiday in honor of the god Saturn, held on the 17th of December of the Julian calendar and later expanded with festivities until the 19th of December. By the 1st century BC., the celebration had been extended until the 23rd of December, for a total of seven days of festivities. The holiday was celebrated with a sacrifice at the Temple of Saturn, in the Roman Forum, and a public banquet, followed by private gift-giving, continual partying, and a carnival atmosphere that overturned Roman social norms: gambling was permitted, and masters provided table service for their

> slaves as it was seen as a time of liberty for both slaves and freedmen alike."

The Catholic Church jumped on the idea of calling it Jesus' birthday beginning the last week of the Roman year in order to lure the Greco-Roman population away from the celebration of Saturnalia and into the Church's version of Christianity.

> "*The Brother of Jesus*" (Hershel Shanks & Ben Witherton III), the discovery in Jerusalem of the possible bones of "James the brother of Jesus" prompted a discussion among the Christian community in which many of them did not realize Jesus had brothers and sisters even though two of the Gospels says that he did and mention his brothers by name…"

That is because of the alleged virgin birth, "immaculate conception," that those same two Gospels have very different accounts of but not found anywhere else in the New Testament.

> "(*continued*) Orthodox church doctrine declared that Mother Mary continued to be a virgin after giving birth to Jesus. So, it has been assumed that Jesus did not have any brothers, older or younger, or possibly that the brothers were from a different mother than Jesus. Both accounts of the names of the brothers are in chronological order with James being first, implicating that James was at least the oldest of Jesus's brothers.
>
> Scholars point out many inconsistencies with the 2 extremely different accounts of Jesus's birth and the

> seemingly fictitious and mythological nature of the story's narrative from both gospel accounts. Moreover, when the book mentions the question that came up early on of whether Jesus was born of Joseph, a paragraph adds in parentheses, (as the Nazareth Jewish community at the time seems to have assumed)"

"Later church doctrine holds" that Mary the mother of Jesus was perpetually a virgin. Never having sex with a man. The Bible itself refutes that assumption by proclaiming in the gospels that Jesus did had brothers and sisters. The situation in Christianity for many centuries was that only the hierarchy and the priesthood could even look at the Latin copies of the Bible that was put together by the order of a Roman emperor toward the end of the 4th century.

So, we see that the tradition of believing Jesus was born of a virgin did not come from all the New Testament writings, only those two gospels and they don't agree on almost any of the pertinent details. In fact, the alleged prophecy of a Messianic figure being born of a "virgin" is vague enough as it is, and now we know that was a mistake in translation by the Greeks who translated the Old Testament before the common era. (The word "virgin" in Greek from the passage of the alleged prophesy was used to translate the word that actually meant young lady or young maiden, but it was not the Aramaic word for virgin.)

Taking into account, if the New Testament says that Jesus had brothers and sisters but church doctrine has been saying that his mother Mary never had sex in her life, because of the bogus immaculate conception of Jesus — and the early Jewish community of Nazareth as well as the Jewish Christians of that era always assumed that Joseph was the father of Jesus — then add to that the inconsistencies and incredibly bogus nature of those 2 Gospel accounts of the birth of Jesus; then one has to strongly consider that the whole story of Jesus's birth is completely fabricated. And here is the main reason why; during the latter part of the first century, after the Jewish Christians had dispersed to escape the war, the biggest issue among the different Christologies that were forming around the stories being told about Jesus was how divine he was, was he always like that, or when did he become that way.

The mythologies of the Greek community at that time is well known. They believed in many kinds of gods and goddesses who could do all kinds of things including procreate with humans. So, in that sense, it was not out of the question in the minds of the Greeks in the Roman Empire that Jesus was born from a god. And the Christology of the Greeks were mostly followers of Apostle Paul, who did not care about or know much about the life of Jesus, including how he was divine.

In fact, there is evidence in Paul's writings that he could have assumed Jesus was an angel before being born. He did at least compare Jesus to an angel. Whether that was just an exaggeration or his actual opinion, we don't know. Moreover, if you look at what the early Jewish Christian community believed about Paul/Saul, it was definitely that Apostle Paul was a heretic who persecuted them and that Jesus was not born divine but was adopted by God in a process from baptism to death to be His Son. The process would have began through the baptism of Jesus and finished when he supposedly resurrected from the grave. The virgin birth was very much a false narrative to make it appear that Jesus was part divine from birth in opposition to other Christological beliefs and the Jewish Christology of Jesus followers who all said that Jesus was fully human, was impregnated by God upon baptism, and was finally born of God when he was risen from the dead.

In the earliest documents of the Gospel of Luke where Jesus is being baptized it says something different than the official version we have in Bibles today because it was changed at some point to correspond with the other accounts. In the original version of Luke's gospel, it says that when Jesus is baptized a voice from the sky includes the sentence, "Today, I have begotten you."

40

Today I have Begotten You

1 CORINTHIANS 15:44-45 (NRSV— written by Paul himself):

> "It is sown in weakness, it is raised in power. It is sown a physical body; it is raised a spiritual body. If there is a physical body, there is also a spiritual body. Thus it is written, "The first man, Adam, became a living being"; the last Adam became a life-giving spirit."

REVELATION 1:5 (RSV), "and from Jesus Christ the faithful witness, the firstborn of the dead."

PSALMS 89:17 (RSV), "And I will make him the first-born, the highest of the kings of the earth."

ROMANS 8:28-29 (RSV), "who are called according to his purpose. For those whom he foreknew he also predestined to be conformed to the image of His Son, in order that he might be the first-born among many brethren."

COLOSSIANS 1:15,18 (RSV— not actually written by Paul):

> "He is the image of the invisible God, the first-born of all creation… He is the head of the body, the church, he is the beginning, the first-born from the dead."

According to the remaining Jewish Christian groups of historical significance, Jesus was adopted by God to be the firstborn, first to be literally born of God. He was not born of a virgin or given a divine spirit at his baptism, that was a seed-form of God from the tree of life which he carried to term upon his death.

What is also fascinating to me is the fact that a definitely real person, James the brother of Jesus, immediately became the head of the Jerusalem Church after the crucifixion even though the Gospels say that his brothers did not support him and fabricated a story that the disciple — whose real name was Simon but was mysteriously changed to Peter — was in charge of the church after Jesus was crucified. It is a well known fact among scholars that this was hidden from the Gospel-reading public of the late 1st century when Jews were at war with Rome.

Orthodox Church history and the synoptic Gospels says Peter was the first "bishop." They deceptively manipulated a passage in the Gospels having Jesus (a Jew) tell Simon (a Jew) that he will be the foundation of his church and his name would be changed thereafter to "Peter" (a Greek name). The Gospel writers were not the people that Christianity has been saying they were. It was not a Matthew, Mark, Luke, or John who wrote any of it, which we know now with certainty. The Gospels came from antisemitic Greek-speaking and writing citizens of the Roman Empire who had been early followers of Apostle Paul's Christology, written many decades after the events they portray by over 50 years or so. With the exception of Apostle Paul and seven of his actual letters in the New Testament, none of the authors of the books of the New Testament were by the people being claimed to have wrote them. It was actually Greeks who followed Apostle Paul during the many decades that followed the events of the crucifixion, particularly after Roman soldiers came down to Jerusalem to put down the revolt that started what became known as the Jewish/Roman War in 66 CE.

The Jewish Christians, the remnants of the original Jesus movement, were dispersed for being messianic, apocalyptic, and very anti-Roman. When Jesus is quoted saying that the kingdom of God was at hand he was talking about a messianic apocalyptic event to defeat the Romans in battle and return Judea back to Israeli control. This made all Jews either hide their identities or leave Judea as Roman soldiers were there to stop any further revolt against Roman authority. It is actually quite crazy that all four Gospel writers made Pontius Pilate look like an advocate for Jesus when the opposite is true and many Jews were executed by Pilate, without a proper trial. (More to come on that subject)

The Roman Catholic organization, an authoritarian, militant, and universal government with a doctrine of taking over the world by force in the centuries that followed, set out to conquer because they believed their Roman based Church was the kingdom of God on Earth with doctrines that were fashioned after the beliefs of Apostle Paul, which was far from anything Jesus taught. Roman Church leaders in the following decades and centuries enlisted what was called "heresy hunters" to seek out and destroy any writings that were contrary to the Roman Catholic beliefs. That included any Jewish Christian writings. We can now understand some of the parables in relation to the tree of life being the instrument for giving a human being the seed of God Himself as was the doctrine of Jewish Christians and their belief that Jesus was fully human before being impregnated and endowed with the Holy Spirit which came from the tree of life when he was baptized.

One of those parables is as follows:

MARK 4:26-29, English Standard Version,

> "And he said, "The kingdom of God is as if a man should scatter seed on the ground. He sleeps and rises night and day, and the seed sprouts and grows; he knows not how. The earth produces by itself, first the blade, then the ear, then the full grain in the ear. But when the grain is ripe, at once he puts in the sickle, because the harvest has come."

MATTHEW 13:2-13 That same day Jesus went out of the house and sat by the lake. Such large crowds gathered around him that he got into a boat and sat in it, while all the people stood on the shore. Then he told them many things in parables, saying: "A farmer went out to sow his seed.

As he was scattering the seed, some fell along the path, and the birds came and ate it up. Some fell on rocky places, where it did not have much soil. It sprang up quickly, because the soil was shallow. But when the sun came up, the plants were scorched, and they withered because they had no root. Other seed fell among thorns, which grew up and choked the plants. Still other seed fell on good soil, where it produced a crop — a hundred, sixty or thirty times what was sown. Whoever has ears, let them hear."

"The disciples came to him and asked, "Why do you speak to the people in parables?" He replied, "Because the knowledge of the secrets of the kingdom of heaven has been given to you, but not to them. Whoever has will be given more, and they will have an abundance. Whoever does not have, even what they have will be taken from them."

As you will see in the following chapters, this information is critical to understanding what the story of Adam and Eve really means. The parable of the weeds and the parables of seeds and trees can be understood with the proper interpretation of the Adam and Eve story after the mistakes in the narrative by the scribes are corrected. There are more parables about seeds and trees than any other subject. That is not a coincidence. Jesus, the human being, believed he had a seed of God inside of him and should carry himself wisely knowing that. Adam and Eve messed up by disobeying God's command to them not to eat the fruit of the tree of the knowledge of good and evil, so now Jesus has to be careful not to mess up like they did.

He was also from the House of David and should have been like royalty but instead he was a carpenter's son living under

Roman occupation. He should be more than that. And now upon baptism, after serious discussions on what he was attempting to do with John the Baptist, he has been awarded this embryo of God inside of him. What does he do? Exactly what he did. His political ideology would be Messianic in that he believed in the prophecies of a messiah coming to defeat their enemies and turn the Holy Land into a garden for him and his followers ruling with the power of his "Father" at their disposal in preparation for the general Resurrection, the kingdom of God on earth. He may have suspected that he would be that Messiah.

There's a good chance that's why he caused the scene at the temple of disrupting businesses there, if that even happened. He could have been trying to portray himself as a leader of the people. There is even a possibility that he did not actually cause the scene in the Temple that the Gospels say he did. The agenda of the Gospel writers was to take away any blame for Jesus' death from the Romans, who were at war with the Jewish population and anyone helping their cause. We know that Pontius Pilate was not the good guy that the Gospels portray, so the agenda of the Greek writers of the Gospels was to divert the blame for Jesus' death away from Romans with the fictitious story of Pilate advocating for Jesus' release. There was never a ceremony of letting a prisoner go. That was completely made up to promote antisemitism. Moreover, everything about what Jesus allegedly did to get himself killed was more political than religious, especially if you take out the miracles and realize that his family could have come from the lineage of King David. Scholars know that much of what ended up being in the New Testament of the actions and words of Jesus are fictitious in origin. Scholars are closer than ever to figuring out what parts of what Jesus did and said are more likely. The opening chapter of this portion of the book by Geoffrey Riggs talks about the methods that scholars are using to determine what a real Jesus may have said. However, context is important. Finding out what exactly Jesus thought and did and why will open up a deeper understanding of what it all means.

41

God Created Man From Ape

Here is my remarkable interpretation of the Genesis story of Adam and Eve that Jesus may have believed and made his mission to fulfill. As fictitious as the tale of Adam and Eve is the way it was written, and the very remote possibility that Jesus of Nazareth (Yeshua) was resurrected from the dead, this may, I repeat MAY have allowed a silent, apathetic, and inactive Designer of the Universe to demonstrate to humanity the purpose for our existence. Revealing the what, how, and why for the existence of the universe.

Ask a theist — one who believes in the existence of a "God" mostly associated with the 3 major religions — what the reason is that "God" created the universe and you'll usually get a blank stare. Perhaps the reply you will receive to this query is that "God" created everything in order to create us, "His children."

First and foremost, we are clearly NOT a god's literal children. However, how would a non-corporeal, not material (not physical), from a higher dimensional plane of existence form of conscious intelligent life procreate? I know it's a stretch to even consider that this Designer of the universe is going to procreate Itself through us. However, what if that was what Jesus thought when he was baptized? To make the offspring of the Designer of the universe through the process of evolution itself with a seed from a tree is not as crazy as it sounds. Living things would not be living if not for procreating it's species generation after generation. Procreation is generally associated with planting seeds. Is it possible that the purpose of everything is to produce actual offspring from what amounts to a deity of some kind, or a species from a higher dimension, or maybe a highly advanced ET? Or is IT evolving us through the natural process of evolution to as high in complexity as humans can become? The universe made all life from itself including us. If a Designer Deity is behind it, what would be the most logical purpose? We are and have been evolving in complexity. How complex a living thing could we become if we continue evolving well into the future and how would that manifest itself? The sky's the limit… Literally.

This theory jumps out of the pages of the Genesis parable and explains much of what Jesus was quoted saying and doing upon understanding the proper interpretation of the story of Adam and Eve in the garden of Eden. It's also how early Jewish Christians viewed what Jesus did and why before fading into historical obscurity. What Adam and Eve was supposed to do in order to "make man in Our image," Jesus may have intended to accomplish himself. Could that describe a process that enables certain individuals in the future to transform into a much more advanced life-form or a newly born offspring of the Designer? Once the proper interpretation of the Adam and Eve story is understood, the possibilities suddenly speaks for itself. It makes sense, and even the parts that seemed in the past to be nonsensical will become logical, practical, and even scientifically cohesive.

Now is the point where the reader must pay close attention to the following texts. The religion of IT is very much dependent on understanding the interpretation of the tale, story, parable, whatever you want to call it, of Adam and Eve.

At the same time, this version of the story directly implies that we did evolve from primates and that other people were around when Adam and Eve supposedly existed. Scholars will say that the confusion comes from there being more than one author of the first chapters of Genesis. No one seems to be aware of the following interpretation of the parable of Adam and Eve in the garden of Eden.

If we are already the literal children of a supernatural entity, then we should be able to move mountains, part a sea, heal the sick, walk on water, raise the dead, or even create a universe. To clarify this logic, I would add; as his "children" wouldn't we consist of the same exact essence, the form of life, or life-form, as His own?

The offspring of any life-form would be the same as that life-form. Right? It would be born as the literal image and likeness. The only rational resolution to this logic is that we are in fact not "His" children in any way, shape, or form at this time and you will see evidence of that in the Bible itself.

Was it the purpose of the universe to evolve physical life into a life-form that the Designer could procreate through at some stage of our development from animals?

This next quotation here implies that one must go through some kind of process to "become children of God." This is relevant to a deist evolution theory of how humans will be the vessels to which a Designer of the universe will procreate through or initiate a process of transmutation to a higher life-form than we presently are, but in the future.

JOHN 1:12-13, (NRSV) "But to all who received him, who believed in his name, he gave power to become children of God, who were born, not of blood or of the will of the flesh or of the will of man, but of God."

Scholars are critical of this Gospel for many reasons. It was written at a time that was too long after the people in the story had all passed away, in an area of Greece where several versions of Christologies were commonplace, and it has statements that would have come from a Christology associated with one called "Gnostic" — who believed that Jesus was part of "God" during the "Creation" event in the beginning of Genesis. The Gospel of John also has some of most antisemitic rhetoric of the New Testament.

We had competing Christologies all over the map concerning who Jesus was in that late first century and early second century era. Some Gnostics believed he was part of God from "in the beginning" and still others believed that the God of Jesus was not the God of the Jews at all. But if you look at what the earliest followers believed, it was that Jesus was a real human being who was impregnated by God at his baptism, although not actually born of God until his alleged resurrection.

The Bible's core beliefs and the religious interpretations of it throughout history are all based on a specific interpretation of the story of Adam and Eve's encounter with two trees in the "garden of Eden." They were allowed to eat from one but not the other. When they disobeyed the rule and ate the fruit of the one they weren't suppose to they were kicked out of the garden and denied further access to the one that they had been allowed to eat from. The true meaning of this tale has been misunderstood ever since Abrahamic religion adopted the pagan belief system that assumes we have an immortal soul of some kind that continues to exist forever even though the Bible itself does not back up that belief.

Keep in mind that these writers did not know that we live on a planet, so everything translated in the whole Bible as "earth" doesn't mean the whole planet. They did not know what a planet is, or the sun, or the moon, or any other celestial body.

According to the interpretation of the Hebrew Scribes who were responsible for the version of the story that was handed down to us, Adam and Eve were the first human beings created and the couple broke an important rule that led to expulsion from the garden with punishments, suffering, and their (and therefore our) eventual deaths. That is how it was interpreted, translated, and copied through the ages. It is full of mistakes.

First of all, in this alternative interpretation we must first build an understanding of what the Bible really says that we are and that allows the Adam and Eve story to speak for itself. To make discoveries in any field of science or scholarly endeavor, one must first do the groundwork. This is no different. So much of what mainstream Christianity teaches as doctrine are filled with mistakes in definitions and interpretations of words and phrases from the Bible. What I was taught about heaven and hell in Sunday School was not at all what the Bible really says once I studied the work of top historian scholars of textual criticism who don't have an agenda to continue the scam that religion has perpetuated.

Religion forcibly interpreted the "immortal soul doctrine" into the Bible. Trying to read something into it that just isn't there.

I am offering this unique interpretation of the tale of "Adam and Eve." I am not familiar with any interpretations like this one by any institution of religion or writer. This interpretation is based on the Bible's own definitions of life and death, not religion's definitions which are based predominantly on the immortality of the soul.

When we evolved from primates, we developed our unique consciousness of self-awareness or possibly acquired it somehow. Medical science cannot find any significant differences between our brains and the brains of our closest animal relatives to explain this. The major difference between us and all other life is that we can openly question the meaning of our existence; to consciously question who and what we are. No other life-form that we know of can do that.

All animals have an instinct of selfishness based on the concept of competition through natural selection. Because we evolved from animals, we carry that mentality of selfishness with us in our DNA and our basic instincts. An animal will not help another unless it somehow benefits itself. A study of chimps, revealed by a Science Channel documentary called "Human/Ape", has found that they cannot understand the concept of selflessness. They only understand competition or cooperation for mutual benefit. It's just part of this physical life-form itself that we prioritize our own survival.

However, we possess the ability to share selflessly as well as the ability to consciously question the meaning of our existence. The three major religions claim that God created us with these abilities when he created Adam and Eve in…

GENESIS 1:26, "Then God said, "Let Us make man in Our image, according to Our likeness."

However, to say "Let Us..." could be to imply that something was/is intended to be accomplished but not necessarily completed at that time. The Scribe who narrated the text proclaims thereafter that humankind "was made" in God's image. Neither does it say that this was when God created "man." It merely indicates that God wanted to accomplish something with "man," how was already here. Without understanding it, they assumed that if God intended to do something, "He" must have accomplished it at that time. But that's not what it says. It says, "Let Us make man in Our image, according to Our Likeness." "Let Us" implies an intention. Nothing more.

If this was the creation of mankind it might have said, "Let us make man and make him in our image." Man was already made, already existing, so the statement "make man in our image" means we already existed.

Here is how it should be interpreted. God was not just intending to make man similar to him. It was His intention to make man into an offspring of His own species. Something Jesus thought he was trying to accomplish. A son of Adam, named Seth, was described in...

GENESIS 5:3, to be "in his [Adam's] own likeness, after his image."

It describes more than just a similar appearance; Seth was his offspring. Just before this verse in verse 1 it says, "In the day that God created man, He made him in the likeness of God."

That verse is one of the reasons that religion has misunderstood the parable. The Scribe who wrote it into this Genesis chapter did not understand the parable and that was a verse added into the text at some point before we ended up with this version. In addition, that passage is part of the Biblical narrative, therefore being just a quote by the human authors or scribal changes (by a Scribe) to the text, not a quote which is claimed to have been spoken by God. All we can say with certainty on this point is that it was only this God's intention to make us in his own image. You'll see the connection coming up.

Webster's Ninth New Collegiate Dictionary definition of "Scribe," "1: one of a learned class in ancient Israel through New Testament times studying the Scriptures and serving as copyists, editors, teachers, and jurists 2 *a:* an official or public secretary or clerk *b:* a copier of manuscripts"

That last verse quoted, 5:1, also reflects the Scribe's Creationist view that Adam and Eve were the first humans to be created and that man was made in God's image on that day, at that time, regardless of what the following verses say about that point. Creationism would most likely be the motivation of the Scribe for adding to the verse "In the day that God created man" to make it say both that God created man on that day and that He made man in his image on that day. The Scribe who added that verse to the parable, whether on his own authority or told to by his hierarchy, did not under the meaning of the parable, in particular, the two trees and what they represented which is fairly well described in the following verses. Also, there are examples in the text coming up of the parable that displays other people being around all over the area. Some scholars will point out that the first chapter of Genesis was an addition to what a previous scribe had written in chapters 2 and 3. But that just solidifies the fact of the influence that the Scribes had on the narrative of Genesis by those who fashioned the text over time to say what they wanted it to say — a creation story and that is all.

[New King James Version (NKJV), old King James Version (KJV), Revised Standard Version (RSV), and New Revised Standard Version (NRSV) throughout; underlining added, and information/comments added within parenthesis or brackets]

Creationism disputes that humankind evolved from animals, believing instead in an instantaneous and magical interpretation of the Biblical narrative in...

GENESIS 2:7 (NKJV), "And the LORD God formed man of the dust of the ground and breathed into his nostrils the breath of life; and man became a living being (KJV, "living soul")."

The logical and correct interpretation of that verse is that we are simply composed of the same materials as everything else here, the dust of the earth, the subatomic particles of atoms, and that we breathe oxygen, the breath of life, to live.

It was not from this verse that we received any kind of "spirit," it was just oxygen, the "breath of life." That is part of the mistaken interpretation by religion.

A "sentient" being like ourselves is one capable of a conscious perception of conscience, having an awareness of oneself as an individual, intelligent and free-thinking life-form. Animals are self-aware to some extent, but that's it. A "sapient" being is "having or showing great wisdom or sound judgment" (dictionary.com). Only humans exhibit both attributes to the extent that we tower over all life in those categories.

We either acquired it or developed it somehow at some point in our evolutionary development from primates. At what point in our evolution could we have changed from non-sapient human/animals to sapient, sentient humans? Is there a clear and logical biblical explanation for it in the form of a parable which has mutated over time by varying interpretations in the story written as "Adam and Eve"?

The New Testament specifically reveals that death is an unconscious form of sleep without dreams awaiting a resurrection. It does not describe a going to heaven or hell mythology. That was forcibly interpreted into the New Testament wherever they could imply it based on the pagan immortal soul doctrine.

If God intended on creating Adam into His own child, His own "image" and likeness, He would have to give him a human "spirit" of self-aware consciousness unique to our physical form which could merge with His own "Spirit" in a kind of impregnation process to grow a seed form of God while being alive. Some kind of merging of temporary spirit with eternal Spirit was the plan in this version of the tale. He would need to give a material thinking being a "spirit" first of our own that could be impregnated with a kind of "seed" of Himself to be grown (someday, somehow) into His literal children. This "seed" of the God Spirit can be symbolized as coming from the "tree of life." Once that is understood, the Adam and Eve tale begins to make sense when interpreted this way.

The Bible uses the word "spirit" in different contexts. For example...

1 CORINTHIANS 2:11, "For the Spirit {"of God"} searches all things, yes, the deep things of God. For what man knows the things of a man except the spirit of the man which is in him? Even so no one knows the things of God except the Spirit of God." (We know that was written by Apostle Paul)

Wherever you find the word "Spirit" in the Bible with the capital "S" it represents God's own Spirit essence. Where you find a lower-case "s" as in "spirit," it is a human spirit. We can recognize the human spirit from anything else because of our awareness of a consciousness inside of us. That is our human spirit.

Apostle Paul only wrote 7 of the 14 writings attributed to him. So, many of the "books" of the New Testament were not written by the people they claim to be or has been assumed to be written by. The best Scholars know that Paul actually wrote only 7 "books" himself in the New Testament. These are the ones that Paul did write...

"Galatians, Romans, 1 Corinthians 2 Corinthians, Philemon, Philippians, and 1 Thessalonians." (Wikipedia or ehrmanblog.org)

All other books of the New Testament alleged to have been written by Apostle Paul are either outright forgeries or disputed over who actually wrote them.

The "tree of life" was one of the trees in the garden of Eden that Adam and Eve were allowed to eat from, at first.

GENESIS 2:9, "And out of the ground the LORD God made every tree that is pleasant to the sight and good for food. The tree of life was also in the midst of the garden, and the tree of the knowledge of good and evil."

Then verses 16-17 proves that God allowed them to eat freely from all the trees, including the tree of life, except for that other one.

"(NKJV) And the LORD God commanded the man, saying, of every tree of the garden you may freely eat; but the tree of the knowledge of good and evil you shall not eat, for in the day that you eat of it you shall surely die."

That should say, "in dying you shall die," The tree of life is a symbolic representation of a seed of God, the "Spirit" of God, the "Holy Spirit" as it is called when it was in Jesus. It does not represent the human form of "spirit," the "spirit which is in man."

Adam and Eve freely ate from the tree of life until they disobeyed God's command to them not to eat from "the tree of the knowledge of good and evil." Every tree produces some kind of seed of itself to grow into an image and likeness of itself. Right? Much of Jesus' "parables" involved seeds and trees. If we have established that God formed mankind from the "earth," gives us the oxygen we need to breathe ("the breath of life") and that the "tree of life" is God's Spirit, or essence, and produces God's own "seed"; how did God supposedly give us some kind of temporary and limited human "spirit," that conscious awareness of self that is necessary to make us who we are and necessary in order to be impregnated by God from the tree of life? When the word "soul" comes up in the New Test it usually means the living breathing body and human spirit together. If we have a spirit of some kind, there is abundant evidence throughout the whole Bible that it is not immortal and all.

Read this closely and attentively. What if the "forbidden fruit", "the tree of the knowledge of good and evil," symbolically or otherwise, contained the "seed" of the human spirit — our unique consciousness? There is evidence in the tale of Adam and Eve that proves how the incident with the forbidden fruit should really be interpreted. I can even provide evidence that Jesus and those around him understood it this way and that Adam and Eve failed to become in the image and likeness of God by having the tree of life taken away from their access. Moreover, Jesus could have easily believed that he accomplished what they failed to do all the way to his actual death. Apostle Paul said it as plainly as I will demonstrate it. From the previous chapter of this book, "Today I Have Begotten You," this was the first biblical verse that I listed…

1 CORINTHIANS 15:44-45 (NRSV— written by Paul himself), "It is sown in weakness, it is raised in power. It is sown a physical body; it is raised a spiritual body. If there is a physical body, there is also a spiritual body. Thus it is written, "The first man, Adam, became a living being"; the last Adam became a life-giving spirit."

Jesus was that life-giving spirit because of the tree of life.

If Adam and Eve had evolved from animals, or was magically created for that matter, they did not have the human spirit until they ate from the tree of the knowledge of good and evil. While they were eating freely from the "tree of life" without a human spirit they would not be impregnated with God's own Spirit, that "seed" of the tree of life. I believe that Jesus of Nazareth understood this and believed he received the seed of the tree of life upon being baptized to grow within his human spirit. He was close with John the Baptist and it was upon his baptism that the main portion of his story began. Convincing evidence of this interpretation is coming up.

Whether the human spirit exists or not is not the focus of this part of the investigation. How a God could even give us a temporary form of his essence is not relevant at this point. Because, since the Bible claims that it does exist, we must look at the Bible to see what it really says according to these ancient texts and common-sense logic.

Before we move forward, the following look at some biblical verses is a quick lesson in what it says on the subject of what we are.

GENESIS 6:17, {the LORD speaking to Noah before the flood} "And behold, I Myself am bringing floodwater on the earth, to destroy from under heaven all flesh in which is the breath of life; everything that is on the earth shall die."

Obviously, he was talking about all oxygen breathing life. He breathed into their nostrils, the breath of life, and they became living souls.

PSALM 104:29, "You hide Your face, they are troubled; You take away their breath, they die and return to their dust..."

ECCLESIASTES 3:18-20, "I said in my heart with regard to the sons of men that God is testing them to show them that they are but beasts. ["beasts", referring to animals] For the fate of the sons of men and the fate of the beasts is the same; as one dies, so dies the other. They all have the same breath."

What happens to our human spirits when we die, and our bodies go back to dust?

ECCLESIASTES 12:7, "And the dust returns to the earth as it was, and the spirit returns to God who gave it."

In this interpretation of the Adam and Eve parable, they took the human form of "spirit" without permission to do so and before they were mature enough to receive it reflected in the act of foolishly putting their lives at risk just to eat something that they were told was poisonous. They committed a pathetically selfish and stupid act. They lived with all of the "trees that bear fruit and is good for food," except for one that is poisonous, because it will kill you. If that was a test to see their maturity and wisdom, they failed utterly."What Adam failed to do..."

Are we aware of anything after our bodies are dead and our human spirits, without a body, returns to God?

ECCLESIASTES 9:10, "Whatever your hand finds to do, do it with [all] you might; for there is no work or device or knowledge or wisdom in the grave where you are going."

Does the New Testament describe our deaths as a form of sleep? Yes.

MATTHEW 9:24, "he [Jesus] said, 'Depart; for the girl is not dead but sleeping.'"

1 CORINTHIANS 15:20, "But in fact Christ has been raised from the dead, the first-fruits of those who have fallen asleep."

There are many other passages that depict death as a form of sleep. This could imply that our human spirits are like rolls of holographic film of an advanced holographic program. Our brains record our lives to our human spirits, like the camera recording the film. When our bodies die, the film stops recording and is stored away (like on a 2-dimensional template) for a possible "resurrection" someday when our universal holographic program is rewound like a roll of film and adjusted for the purpose of a resurrection from the dead, possibly to take place in a higher dimensional plane or at the end of time. Ancient people would find it exceedingly difficult, if not impossible, to understand these concepts. This investigation in Quantum holographic cosmology will provide a growing awareness in the finest minds of theoretical physics toward answering the deepest mysteries of the universe. Then the convergence of science and religion will be inevitable. After the following definition will be the evidence of this interpretation that I've promised.

Webster's Ninth New Collegiate Dictionary definition of consciousness;

> "1 A: the quality or state of being aware esp. of something within oneself B: the state of fact of being conscious of an external object, state, or fact C: AWARENESS:
> esp concern for social or political cause 2 : the state of being characterized by sensation, emotion, volition,

and thought : MIND 3 : the totality of conscious states of an individual 4 : the normal state of conscious life 5: the upper level of mental life of which the person is aware as contrasted with unconscious processes"

The Bible says that Adam and Eve were "naked and not ashamed" until the moment that they ate from the "tree of the knowledge of good and evil."

GENESIS 2:25, "And they were both naked, the man and his wife, and were not ashamed."

This was before they ate the forbidden fruit. The moment that they did however is the moment that they acquired the human form of spirit that represents our consciousness.

GENESIS 3:7, "Then the eyes of them were opened, and they knew that they were naked; and they sewed fig leaves together and made themselves coverings."

In verses 10 and 11 God confronts them in which Adam said to Him…

"I heard your voice in the garden, and I was afraid because I was naked; and I hid myself."

The very next thing God says is...

"Who told you that you were naked? Have you eaten from the tree of which I commanded you that you should not eat?"

Careful consideration of that passage reveals a definite and undeniable connection between eating the fruit of the "tree of the knowledge of good and evil" with the awareness of being naked. At this very moment, they were no longer just highly evolved primate, human-animal life-forms — whether a real incident or symbolic — they now had the "spirit of the man which is in him," the "Higher-self" (mistaken by some of religion to be our "soul"); and having the ability to merge with the God "Spirit" to plant His "seed" into this human "spirit." Adam and Eve went from highly evolved human animals from primates to beings with a "quality or state of being aware esp. of something within oneself." Or, "the upper level of mental life of which the person is aware as contrasted with unconscious processes."

In other words, a conscious awareness of the self inside of us, a consciousness.

This is what a historical Jesus likely believed as records of early Jewish Christians and parts of the Gospels and Epistles of Paul can confirm by putting it all together like a jigsaw puzzle. When Jesus was baptized and began his mission, he was under the impression that he literally walked with God. The Spirit of God from the tree of life was merged with his human spirit. Luke's original version of Jesus' baptism has God saying, "Today I have begotten you." In other words, at that moment is when Jesus was "filled" with the Holy Spirit, the seed from the tree of life. Here comes more undeniable evidence.

This act of Adam and Eve receiving the human spirit at that moment made them become consciously self-aware, separating humans from animals with our conscious awareness of the self. This becomes clear with the next verses. After God tells the serpent, Eve, and Adam what the consequences will be of their actions...

GENESIS 3:21-22, "Also for Adam and his wife the LORD God made tunics of skin and clothed them. Then the LORD God said, "Behold, the man has become like one of us, to know good and evil. And now, lest he put out his hand and take also of the tree of life, and eat, and live forever — therefore the LORD God sent him out of the garden of Eden…"

That was to prevent Adam from having further access to that tree of life (Holy Spirit) and prevent him from growing that seed from the tree of life into an offspring of the God family, "like one of us." That would be the purpose of everything in a biblical hypothesis. The Religion of IT! Is the universe is a creation for the evolution of humankind into God's children? Maybe so, I don't know.

Now Adam became like one of them, a being with a self-aware consciousness, and a "spirit" capable of impregnation by the God Spirit. A life-form with a conscious self-awareness of intelligent thought who could now use the tree of life to live forever.

Now, if they also continued to take from the tree of life, they would receive the Seed, Holy Spirit, "image" of God and could become an eternal ("live forever") offspring of God once the body dies and the seed sprouts into IT's offspring, the image and likeness. If a human Jesus thought that the baptism ritual gave him access to that fruit of the tree of life, he would believe that he must grow into an embryo of his God and maybe become born literally of that God after his death when the seed decays into the soil and sprouts a new tree. The process of impregnation of the human spirit with a seed form of the Designer from a "tree?" It's a possible interpretation. God stopped this process from happening to Adam and Eve by taking the tree of life away and banishing them from the garden of Eden.

If a God exists, this is the most logical reason possible for creating this universe, to procreate an inter-dimensional life-form. Every living thing is born and must grow and evolve within this 3-dimensional {time is the fourth} holographic material that we call the space-time continuum that Stephen Hawking called a "biofriendly, living organism."

On this interpretation of the parable, we can speculate even further on its meaning in reference to this deist evolution theory and the evolution of the human consciousness.

However, why did God prevent Adam and Eve from becoming His Children? Because God had told them not to eat something, something deadly, but they did anyway.

Whether an actual event or just a "parable," there could be a process for us to undertake in order to advance in more complexity or to become His Children, to "live forever," which is the only logical purpose any intelligent entity would have for making this universe in the first place.

What I have laid out as scientific discovery in PART ONE is compatible with this religious theory. A process that the characters of Adam and Eve failed in our first Test, but necessary in the development of Humankind! To become sentient human beings from evolved human animals for the purpose of evolving into a higher life-form. Or at least some kind of metaphysical connection with the mysterious and as yet completely unknown Impersonal Universal Consciousness that I very much expect is our Designer of the universe.

This clarifies that now Adam and Eve obtained a "spirit" which could merge with God's "Spirit" to be grown into a literal child of God. When they were eating "freely" from the "tree of life" they did not yet have this human "spirit." The "tree of life" had given them some kind of metaphysical connection to God's "Spirit," but they did not yet have the human spirit for the growing process to begin. Without the human "spirit" from the "tree of the knowledge of good and evil" they could not "live forever" even though they were eating "freely" from the tree of life. Now that they had taken the human spirit without permission, God stopped the process. In verses 23-24 God placed an angel to "guard the way to the tree of life" with a "flaming sword."

The "tree of the knowledge of good and evil" was that human spirit, whether it's real or just symbolism. But by taking this for themselves, against God's "Command" to them not to, they were NOT tortured in flames or killed, they were simply cut off from the "image and likeness" of God, the "Holy Spirit," the "tree of life," the "kingdom of God." What truly is a "kingdom" if not a species of life and/or a ruling family? The idea that a "kingdom" is just a place or political system has limited our perception of its possible potential meaning here. Only the Gospel called "Matthew" refers to a "kingdom of heaven" while the rest of the Gospels refers to it as a "kingdom of God." The reason for that was Matthew's desire not to offend fundamentalist Jewish people by using God's name. So instead of saying kingdom of God he frequently wrote kingdom of heaven. They believed that God's throne was in a "heaven" which at first just meant everything above their heads. So, a place where God is located to them would be described as a kingdom in, owned by, ruled by and of "heaven." The following chapters goes over the details of what the Bible really says about the words heaven and hell.

According to this interpretation of the Adam and Eve parable, the true evolution of humanity started from the Animal Kingdom and became the Human Kingdom at that moment; for the overall purpose of joining the God Kingdom. At some point before humanoids developed a conscious and sapient sense of self-awareness, they had acted like animals in the sense that they didn't cover up their private parts.

Adam and Eve did not have that refined ability which separates us from animals, our innate conscious perception of self-awareness, until "their eyes were opened, and they knew that they were naked." This interpretation could have been the original one thousands of years ago or it just happened to complete the puzzle by the mere chance that it could. Maybe so, I don't know.

How long ago did our ancestors transform from being just animals to becoming self-aware, sapient, and sentient? Does this parable imply that we evolved from being just animals into something entirely unique in this material universe and a further stage in the evolution of life toward the Designer's overall goal? This theory flies in the face of Darwinism, Creationism, and intelligent design.

I've wondered for many years now why neither Jews nor Christians have made this interpretation from the Bible's Adam and Eve story. As we investigate the Bible further to see that there were other humans existing at the same time as Adam and Eve, you will see how this jigsaw puzzle comes together to paint a fairly clear picture. One of the deceptions that religions continue to teach is that the human "spirit," also mistakenly portrayed as the "soul," is either the "image and likeness of God," and/or already immortal {"live forever"} self-existent life and/or already created as God's "children." Some will claim that the human spirit itself is the "image and likeness of God" without any evidence of that in the Bible. Much of Christianity claims that it is and claims that it was given to Adam in the instant that man was created from the dust of the earth. That claim depends on the indoctrinated belief in the supernatural and further distorts and hides this alternative theory and interpretation of what we are and why we, and the universe itself, exists. This is by far the most logical way to look at it.

When I told a Jewish friend my theory that our purpose is to become God's children, his response was to say that "We are God's children." Which I followed by just saying, "Not literally!" While they hold on to believing that those archaic fables and parables are literal and historical records they refuse to accept that humanity evolved from the Animal

Kingdom. The truth is, of course, that we did, and the Bible can confirm that hypothesis.

We have a material-based life that depends, thinks, and acts on material instincts inherited in all life for the enhancement of our survival instinct itself. Any eternal, inter-dimensional, and self-existent life-form does not need our survival instincts to exist. IT does not need or want anything and therefore would not need, want, or desire at all to be selfish. Why would it behave and think like us temporary physically bound human beings? Taking the forbidden fruit just because we wanted it was a blatant selfish act. The selfish mindset is completely contrary and antithesis of the kind of mindset that would be necessary to grow the "seed" of the "tree of life." Jesus tried to teach that.

This understanding is why Jesus was so benevolent. He knew this. This world now, under mankind's control with capitalism, would be almost impossible to grow this God-seed. It's even possible that this "God" (if "He" is the Designer) chose Jesus, "Yeshua," to be the "firstborn"; and so far, the only one — even though it is not likely that Jesus resurrected from the dead… at least not yet anyway. And note that Jesus believed that he had this "Holy Spirit" and lived with it in the world back then for 3 years until his death. It was his possible belief that he was the prophesied Messiah or from Jewish royalty ("THE KING OF THE JEWS" was written as a "notice" near his cross according to the Gospels because of his family supposedly being from the house of David) got him executed for sedition or insurrectionist activity against the Roman occupied authority.

PSALM 2:7, (NRSVUE) "I will tell of the decree of the Lord: He said to me, "You are my son; today I have begotten you.""

Imagine what would be the first reaction of an animal if suddenly given our kind of consciousness, the perception of self-awareness? Its eyes would become wide open, consciously self-aware, "the upper level of mental life of which the person is aware as contrasted with unconscious processes," and it would know that it was naked. This was Adam and Eve's reaction once they ate from the tree of the knowledge of good and evil. It all fits together and is perfectly logical. As I have said, if facts have logic backing it up, it can't be easily refuted.

Adam and Eve were not capable of or having the potential for eternal life; until they received the human spirit, which can be merged with God's own Spirit to be grown into the exact image and likeness of God, which Jesus and Apostle Paul thought that he accomplished when he died. This is why Adam and Eve would not have lived forever before they ate the forbidden fruit even though they were eating freely from the tree of life. They did not yet have the human spirit until after they ate the forbidden fruit. It is logical and makes perfect sense if you let go of indoctrination and think it our clearly.

Once those narrative views in the text that don't make sense are eliminated, the parable reveals it's true potential meaning. By believing the narrative of the text, it looks like God was lying to Adam. So, people who are indoctrinated are tricked into believing what the serpent said that God was lying. Right?

If you believe what the text of Genesis itself as written by the Scribes says about it, you are saying that God lied to Adam, "in the day that you eat of it you shall die." But they didn't die, so the serpent was telling Eve the truth and God was lying? That's the way the text reads if you don't see the bigger picture and replace that verse with "in dying you shall die."

There is evidence that religion has never been able to logically explain based on their stubborn insistence that Adam and Eve were the very first human beings. After Cain, their son, was outcast for killing his brother Abel, he married a woman, had children, and built a city.

GENESIS 4:17, "And Cain knew [a biblical term for sex] his wife, and she conceived and bore Enoch. And he built a city and called the name of the city after the name of his son — Enoch."

If religion's interpretation would have us believe that humankind did not evolve from the same ancestors as apes, and that there were no other human beings here before Adam and Eve were created from the dust of the world, where did Cain's wife come from and the people who built and inhabited the city? It is illogical enough to make biblical interpretations which completely ignore and contradict evidence from the sciences like physics, evolutionary biology, astrobiology, archaeology, anthropology, and paleontology. It is even more illogical to attempt explanations for how the Bible itself refutes their own doctrines, interpretations, and beliefs.

Here is even more evidence of other people around when Adam and Eve existed according to Genesis. When God put a curse on Cain for killing Abel, Cain says to God in...

GENESIS 4:14-15, "Surely You have driven me out this day from the face of the ground; I shall be hidden from your face; I shall be a fugitive and a vagabond on the earth, and it will happen that anyone who finds me will kill me.' And the LORD set a mark on Cain, lest anyone finding him should ["would" or "could"] kill him."

If Adam and Eve were the first human beings, there would not be other people, "anyone," around who might try to kill Cain. Christianity has never been able to deal with these contradictions in what they believe. It also means that this original God character didn't put Cain to death for killing Abel but rather put a mark on him to keep him safe from others. That doesn't sound like the vengeful authoritarian God of the commandments, genocides, and baby killers of the Old Testament.

There is a design to the universe. Our universe, galaxy, solar system, and planet was suited perfectly for human emergence and habitation because its designed design made it inevitable to happen somewhere in the universe. Every seed grows into the tree that it was designed to become if the timing and conditions are right for it to grow. If the soil is rich and the water and light is in the right amounts, it will grow. Our planet keeps its established distance from the sun, revolving on a particular axis that creates stability and the seasons. The right amount of gravity keeps our feet perfectly on the ground and everything in its place. What kind of life-form could create this universe and why?

A deist design theory like my deist version of designed evolution can reveal some real logic from it. It was all designed to begin with to function naturally as it has with no interventions, guidance, or the direct of indirect involvement by the Universal Consciousness ITself.

GENESIS 3:17-19 (NKJV, God's punishment to Adam) "Cursed is the ground for your sake; In toil you shall eat of it All the days of your life. Both thorns and thistles it shall bring (cause to grow) forth for you, and you shall eat the herb of the field. In the sweat of your face you shall eat bread till you return to the ground, for out of it you were taken; For dust you are, and to dust you shall return."

Imagine this as an embellishment by its human narrators of the consequences of being kicked out of a lush garden. While living in the "garden," God had (Gen. 2:9) "made every tree grow that is pleasant to the sight and good for food." Imagine having every tree that produces "food" provided for you and then taken away? Would it in this case then be a punishment to be forced to provide for yourself or a consequence of your own actions? The primitive narrators felt that it was a punishment and embellished the story to appear that way. Ancient people believed a god, or "God," provided them with everything. If it rained, it was a god or "God" that had provided them the water. It was not a natural occurrence to them; it was God or a god. If they fell and hurt themselves, it was not a consequence of being clumsy, or not looking ahead at where they were going, it was their punishment from God or a god for doing something wrong. Eve's punishment was to feel “pain” during childbirth (Gen. 3:16). However, if they had stayed in the garden under God's care “He” obviously would have alleviated or eliminated that pain. At an earlier point in the narrative, God had brought every animal to Adam for him to see and name (Gen. 2:19). Obviously, God would not have allowed any of the predators to attack Adam or even other animals. Also, note that the punishment to Adam implies that Adam will eat only from “the herb of the field” and “bread.” No mention or implication that he could eat meat, whether clean or unclean, at that point.

In forcing Adam and Eve out of a “garden” and the comforts of being under God's direct care, they would have to fend for themselves now. Not as a punishment, as the biblical authors assumed, but a consequence of their actions. The authors of the whole Bible did not understand this, we can!

Furthermore, Adam and Eve did not exist! That means that there was no “original sin” that they committed that all humanity falls under... Let that sink in... Let the facts plow through the wall of lies and confusion in so many of your minds. Adam and Eve is a parable. It tells a fictitious story that has an important message. They failed a test to qualify to become impregnated with the tree of life. Even further, it did not have anything to do with what religion calls “sin.” That is simply a manufactured primitive concept. It was not even the “sin” that got them eventually killed, it was for not receiving the seed of the tree of life and growing it once they had the human spirit to merge with. You should be able to see why the original sin, the salvation from sin, and "redemption" are all nonsense because it was NOT the reason that Adam and Eve died or the reason that WE die! It is in dying that we will die unless we are somehow, someway, and someday able to receive and grow something inside of us that causes a transmutation to a higher life-form. It brings a logic to it all that has never been fully realized.

The claim that it was the sin that caused Adam and Eve's death penalty is a huge mistake in interpretation going all the way back to the scribes who made scrolls of the story for millennia before the Greeks made the first Greek version of the Hebrew Bible, the Septuagint, during the third century BC.

Death was NOT because of a sin; it was the consequence of breaking that one important rule of common sense (do not eat something deadly) and then being cut off from the tree of life.

That is where the nonsense of human sacrifice and salvation from sin comes from. The death of Adam and Eve, when correctly interpreted, was not from the act of disobedience or from the forbidden fruit itself. It was in losing access to the tree of life because without that, in dying you shall die. It's really that simple.

With no original sin, there is no need for a sacrifice or Savior from sin. No need of a primitive human sacrifice to a god. That eliminates that whole "Jesus died for our sins" doctrine. That which we perceive on faith alone is not credible to common sensed intelligent minds with conscious awareness of self. Let us stop being animals and foolish humans. History has many unfortunate examples of that, and it is still happening today, as you well know.

42

Religion's Exposition

The following is another examination of the proper interpretation of the parable of Adam and Eve. It has significant meaning when understood in its original context.

When human beings evolved from their primate ancestors, they were just animals with more intelligence than others. Acting on instinct and somewhat aware of existing, but not sub-consciously aware. That would allow asking questions like "What am I?" All animals are self-aware to the extent that they know that they exist but are not conscious of it in the same sense that we are. They cannot question it. They don't know they're naked and don't care. Only human beings are consciously aware of being self-aware. Only humans are aware of the fact that we are self-aware and as a result, can ask the why questions.

GENESIS 2:7, (NKJV)"And the LORD God formed man of the dust of the ground, and breathed into his nostrils the breath of life; and man became a living being (KJV "soul")."

It is not the biblical "soul" of a person at all. There's no such thing. The "soul" was a word used by the scribes of Hebrew scripture almost 3000 years ago during phases of writings being copied for preservation. It said in the King James Version that the man became a living soul when the breath of life was breathed into his nostrils. That means that without the breath of life, man becomes a dead soul. He does not have a soul, he is a living soul while he is alive. That's all it really means.

The human spirit from the tree of the knowledge of good and evil would change Adam and Eve from being completely influenced by instinct (like animals) alone to being consciously aware of existence (knowing that we exist and are naked). This seed of the human spirit would also enable us to be impregnated by a seed from the tree of life, referred to by the phrase Holy Spirit and Spirit of God.

"*A History of the End of the world*" by Johnathon Kirsch, pg. 190-191;

> "A plain reading of Revelation suggests that everyone on earth in the end-times; men, women, and children, saints and sinners alike — will be compelled to endure the suffering to be inflicted on humankind by the Antichrist during the final years of persecution and oppression known as the Tribulation. Only after the Tribulation is over will the dead saints and martyrs be raised from the grave... [C]ertain cheerful Christians in nineteenth-century America, however, refused to believe that they would be called upon to endure such afflictions, and they insisted on embracing a new and highly inventive version of the end of the world. Christians who are worthy of salvation, they preferred

> to believe, will be miraculously plucked up and elevated to heaven before the Tribulation begins in earnest. Seated in the galleries of heaven, they will be privileged to look down and watch as everyone who has been left behind on earth suffers and dies under the Antichrist. Only when the Tribulation is over will they return to earth in the company of Jesus Christ to dwell in the millennial kingdom. Their comforting theological innovation came to be called the Rapture.
>
> [N]either the word nor the concept of the Rapture is mentioned anywhere in Revelation. Rather, the whole notion of the Rapture is based on a couple of lines of biblical text in the First Letter to the Thessalonians, the earliest of Paul's writings and perhaps the single oldest document in the New Testament. And, significantly, Paul seems to believe that the remarkable events he describes will take place in his own lifetime rather than at some unknown point in the future."

That is what religion does, invent their doctrines by mixing and matching scriptures and parables, with infallible words written by delusional people who then worship their own made up stuff. They believed that something that was written by the demented lunatic named "Apostle Paul," the enemy of the disciples of Jesus and all Jewish Christians thereafter, was a message to them that they can interpret by saying that they will greet the returning Christ in the air. Nothing that was written in the Holy Bible of Christianity should have ever been taken that seriously. Case in point...

43

The Stars Falling From Heaven

Neil deGrasse Tyson:

> "If you knew nothing about science and read, say, the Bible, the Old Testament, which in Genesis is an account of nature, that's what that is, and I said to you, 'Give me your description of the natural world based only on this [the Bible].' You would say that the world was created in 6 days and that the stars are just little points of light, much lesser than the sun and they can fall out of the sky — you know, one of the signs of the second coming is that the stars will fall out of the sky and land on Earth. So, to even write that means that you don't know what those things are. You have no concept of what the actual universe is. So, everybody who tried to make proclamations about the physical universe based on Bible passages got the wrong answers. So, what happened was, when science discovers things and you want to stay religious, or you want to continue to believe that the Bible is unerring, what you would do is, you would say, 'Well, let me go back to the Bible and reinterpret it.' Then you say things like, 'Oh, they didn't really mean that literally,

they meant that figuratively.' So, this whole sort of reinterpretation of how figurative the poetic passagės of the Bible are, came after science showed that this is not how things unfolded."

MARK 13:25, (ESV) "And the stars will be falling from heaven, and the powers in the heavens will be shaken."

REVELATION 6:13, (ESV) "And the stars of the sky fell to the earth as the fig tree sheds its winter fruit when shaken by a gale."

MATTHEW 24:29-30, (ESV) "Immediately after the tribulation of those days the sun will be darkened, and the moon will not give its light, and the stars will fall from heaven, and the powers of the heavens will be shaken. Then will appear in heaven the sign of the Son of Man, and then all the tribes of the earth will mourn, and they will see the Son of Man coming on the clouds of heaven with power and great glory.

ISAIAH 34:4, (ESV) "All the host of heaven shall rot away, and the skies roll up like a scroll. All their host shall fall, as leaves fall from the vine, like leaves falling from the fig tree."

NOTE: A particular author I quote from the most here is the highly respected and considered by many of being the top expert scholar in the field of New Testament writings, historically, and "textual criticism," of the Holy Bible, Bart D. Ehrman; admitted agnostic, historian, "the James A. Gray Distinguished Professor of Religious Studies at the University of North Carolina, Chapel Hill, and is the author of more than twenty books, including the New York Times bestselling Misquoting Jesus; God's Problem; Jesus Interrupted; and Forged, as well as Did Jesus Exist, How Jesus Became God, and Armageddon.

The occupation of Judea by the Roman Empire was the reason for the revolts, insurrections, Jesus' crucifixion, and the following "Jewish War" against the Roman Empire from 66CE ("Great Revolt") that lingered and boiled over again 50 years later, only to have the practice of Judaism of that era banned from Jerusalem in 135CE. Since Jesus' Christology was much more Jewish than any other type of Christology, the true followers of Jesus were the ones who were forced out of Jerusalem, persecuted by nearly everyone, fed to lions, and slaughtered by gladiators in arenas or by Roman soldiers on the battlefield.

44

The Tree of Life

There's nothing in science or in the Bible that explicitly or precisely says that we are already immortal. That only came about in the minds of human imaginations of mythological doctrines. Jewish indoctrination taught during Jesus's lifetime, that we are resurrected physically someday from the sleep of death or we simply do not exist ever again. The "soul" is not immortal according to nearly every word of both "Testaments" and common sense. It would obviously be a universe of chaos and self-destruction if a million psycho-killers were allowed to live eternally.

On the other hand, the concept of a hell of Greek mythology is a monstrously cruel thing to do to anyone and a ridiculous concept to logical thinking intelligent minds. So, as you will see in the following chapters, the Bible actually teaches that a "Gehenna" fire, one of the three Greek words that was translated in 1611 into English to the word "hell," is described as a "second death" with no further resurrection in which human spirits put into this, or the lake of fire will "perish" (burned up until nothing is left) in flames that will not be "quenched" (put out). Which means, the flames will not be put out before destroying everything that it is meant to destroy. Once it has consumed all that there is to burn up completely, the flames will die down and go out like all fires that are not quenched. It's like any camp-fire that goes out after burning off all the wood available for it to burn.

However, according to records of what the original Jewish followers of Jesus believed and passages from the Old and New Testaments, no one has ever had a second shot at the tree of life except for Jesus when he was baptized (impregnated by God).

If there is any truth in the Bible about a God of everything, the tree of life is symbolic of the overall purpose, the very reason the universe exists. Even though our subatomic particles are interconnected, our human spirits of consciousness are not connected to each other in the same way but will be for those who evolve to a higher plane. Our human spirits came symbolically from the tree of the knowledge of good and evil which are like temporary templates of 2-dimensional holographic films; similar in scope to video footage being recorded of our lives by our minds. But none of it is immortal. The entire Holy Bible and all of our empirical science makes that fact very clear. The immortal soul doctrine was an easy way to put one's mind to rest over the fate of death. If we continue to exist in some place outside of our bodies, they reasoned, there's really nothing to worry about. Especially if a paradise is envisioned to be one of the possible destinations. It's wishful thinking at the most childlike level.

Now when you read the first chapters of Genesis after reading this book, that much will finally make sense. When Adam and Eve ate the forbidden fruit, they instantly became consciously self-aware by realizing that they were naked. They went from evolved animals who didn't care that they were naked to a consciousness of self-awareness that does care. They now contained a spirit form of self-awareness of some kind according to how the tale unfolds. They became consciously aware of being consciously self-aware and naked. And this particular human spirit can also be impregnated by the Holy Spirit from the tree of life to form an embryo of a hybrid life-form of some kind, which Jesus may have believed would make him the first offspring of his God/god. Whether he believed that it actually was the Jewish God that was present inside of his human spirit, we can't know. So much of the Gospels are unreliable. It could have been a different kind of god altogether, even an extraterrestrial. It's the reason Marcion thought it was not the God of the Jews who was the deity that fathered Jesus. They seem to be too different in every way. The passive, selfless visionary of a humanitarian Jesus as opposed to the maniacal, psychopathic, totalitarian, despicable despot of the Jewish God.

The parable simply portrays a primitive description of how Evolution could have taken us in the development of our sentient and sapient minds. It's an explanation for how we acquired a spirit of consciousness that only human beings possess. But this conscious self-awareness is not immortal. The purpose is to obtain access to that enigmatic tree of life and grow a seed of some kind of multi-dimensional, advanced form of life. Any seed does not sprout until the seed itself begins the germination and decomposition process, decaying in the soil. The seed itself must die in order to sprout a sapling of the tree. Right? That is why the Jewish Christians insisted that Jesus did not become divine or born of God until after he was dead. It is the most logical interpretation and backed by what we know the Jewish Christians believed before fading into historical obscurity.

Moreover, it would seem that a real Jesus probably knew this and it was a huge part of his motivation. When he was baptized by John the Baptist, he believed he obtained a literal "seed" of God, and that he was suddenly given what they called the "Holy Spirit." The Greek writers and Pauline Orthodoxy could not allow that interpretation to stay in Luke's Gospel so they changed it. That is another clue that the Virgin birth story is a fable that was written into the Gospels of Luke and Matthew by their authors who had certain agendas to mutate their Christology with ancient Greek mythology. The point of making up the story of the Virgin birth was to say that Jesus had part of God in him from his birth or before. It was an important issue between different groups of Jesus followers for decades after the crucifixion. Especially in opposition to the Jewish Christian claim that Jesus was not divine at all until his death and resurrection. Furthermore, the one that makes more sense is the "Today I have Begotten you," a fully human man taking the seed of the tree of life to grow into a newborn of God, completing the process that Adam was chosen for but failed because of a stupid and selfish decision to eat something that he was told was poisonous.

The mythological and totally fictitious virgin birth version won the war of attrition while the original Jesus Christology faded into obscurity following the Jewish war in the decades and centuries that followed. What ended up as Christianity had no resemblance to the Christology of Jesus and his "Jewish Christology." It is a fact that most Christian scholars prefer to keep secret.

It's obvious now that the two completely different versions of the Virgin birth in the Gospels are so completely different and fabricated that you can't trust any of it to make any sense. If there's one reliable indication of something being untrue without concrete evidence, it is common sense. There is no common-sense or logic in those Virgin birth stories, particularly considering the wide differences between the two versions. It is propaganda of mythological influence in competition with other Christologies. Critical scholars have picked it apart, dissected it, and revealed so much evidence of fabrication, I can't believe that our presently evolving generations still believe any of it. The number of young people becoming involved with religion is dropping with every generation. The more we know, the more we can grow.

45

The Three Hells

A closed mind is a trap! The religious belief of requiring faith to believe is to insist that something is true without evidence or question. When there was no science to explain how our planet works, and explain to us what everything is, humanity used their imaginations to believe it was magic of one kind or another and it was gods — powerful, omnipotent (or so) and intelligent beings, and similar in many ways to us. They didn't create us. We created them. So, we required "faith" to believe the fantasies that explained our existence in ways that we could believe, whether it made sense or not. Authoritarians have always tried to use religion to keep the masses in line and obedient to them. They needed to assume that we were created by beings like us, that we could relate to, in the absence of any knowledge about the world. All gods were made by man in our own image and likeness. Why isn't that obvious to more people? Again, it's indoctrination and a need to believe.

This physical universe is a thing which is not self-aware, sentient, or asking questions. It created us from itself, a life-form (us) that is self-aware, sentient, and asking questions.

To Atheism it was caused by a random occurrence for no reason.

To "Creationism" and intelligent design it is the result of guidance, direction, interventions, and manipulation by an omnipotent, supernatural deity — also for no particular reason. In my hypothesis, this is all part of the natural design of the universe, neither guided nor random, but it was engineered and set in motion by a MIND of some kind who has left us in the dark about what we are and why we exist, for the moment — as we slowly develop into the benevolence of altruism.

If atheism is the correct answer, why would a universe that cannot ask questions create from itself a life-form that can? If the universe is not the result of a designed design by a sentient and sapient mind, why would it create a material being from itself with a sentient and sapient mind? Atheism cannot logically answer that. On the other side, if the correct answer is a guided (intelligent design) universe by a deity, why is there no direct contact between humankind and that deity except what was written by humans so long ago that we can't trust it? The Bible is not written like an instruction book to humanity from their maker, nor any kind of historical record. If the religions of the world require myths, legends, and ancient writings to even guess at the who, what, and why of the Designer of the universe it means that we don't know and whatever "He", She, or IT is doing with this planet, IT has left us completely in the dark. The most logical reason for that is because it was necessary in the process of evolution from animals to keep us in the dark until we're ready. We aren't ready yet.

The extinct closest relative of modern humans, known as "Neanderthal," buried their dead some one hundred thousand years ago. They placed food and tools into the graves of their loved ones, believing (or just hoping) that the items would be needed in the afterlife. From at least that far back, humankind has needed to hope that their lives would somehow continue after death. It is a survival strategy to some, a way to lessen the fear and uncertainty of death to others, and a way to continue living their lives completely on their own terms, imaginations, and delusions with little concern for the future of others or for anyone but themselves and the extensions of themselves, as in natural selection. The immortal soul doctrine and religion in general is a drug that many get themselves addicted to. That wouldn't be a terrible thing alone, but they force everyone to bow to their fictitious gods by making the laws of the land according to their particular choice of scriptures from their infallible sacred writings and interpretations of it.

Our brains are a combination of chemicals, tissue, and electrical impulses which direct all the physical functions and mental capabilities of the person. We are self-animating materials and physical impulses with an advanced form of sentience unlike anything else in our existence. Are we something more than that? We want to think that we are. I am agnostic, so, I don't know, maybe so.

The "New Testament of the Holy Bible" has been translated into other languages from the original Greek that the New Testament was written in and the "Latin" which was the first language that it was translated into. It was only found in those two languages until 400 years ago when it was translated into German, English, and many others thereafter once the printing press was invented. Over the many centuries leading up to that point, all religion was heavily influenced by the popular Roman, Greek, and Egyptian mythologies as well as the first official organized Christianity, Catholic. When scholars got around to translating the Bible into English from the Latin version and some of the original Greek, the popular beliefs of their religion greatly influenced their work. In particular, the beliefs of what we are and what happens after death. As you have seen, and will see proof, it turned out that their belief in an immortal soul that goes to a heaven or hell upon death is not backed up by the Bible. This caused English translators to interpret words and phrases based on the religious beliefs they held, not necessarily what those words and phrases meant in the context of what was written.

Scholars have found reasonable evidence throughout the New Testament that much of it has questionable origins and/or fictitiously embellished narratives and events with erroneous details which changed over time as different copies were copied (by hand you know).

It's all Greek to me...

The old English word "hell" was used by the English scholars around the year 1600 to translate three completely different

Greek words from some documents they had of the "New Testament" texts. There are in existence today few documents preserved of those writings in the original Greek and very many Latin versions but only going back as far as the birth of the Roman "Catholic Church," the Orthodoxy.

All of the documents preserved have differences. No two documents of text are exactly the same. Another problem was how words of all languages change over time. Translating the texts was not a simple cut-and-dry procedure. Complicating the process also was how Christianity held beliefs at that time, and still do, that the documents did not confirm. These chapters make an in-depth study of what Christianity believed at that time, most still do now, and what the surviving texts of those ancient writings actually says. In this chapter, it's about the word "hell."

The top reason there has been so many differences and so much controversy between institutions of Christianity is because so much of their "Doctrines of Beliefs" are interpreted from the Bible by twisting and spinning the meanings of words and phrases into however they wanted to interpret it along with the inadequate way it was translated. There are some Sabbath-keeping forms of Christianity who know some of this and yet continue to believe and observe traditional Christian religious practices and doctrines which are part of the mistakes by translations and interpretations going back through Christian history.

What the Holy Bible plainly says is not what the vast majority of Christianity believes. It is also complicated by the fact that contradictions in the writings point to the confusion of the authors themselves in their own beliefs and the changes made over time by those religious Scribes making the copies centuries later. Until the invention of the printing press, copies had to be made by hand of documents which had been copies made by hand of documents that were also copies made by hand and so-forth going back to the first century. Many changes took place along the way by Christians influenced by many changing factors.

The differences between the documents of the "New Testament" writings available to scholars for examination is so vast now that "Bart D Ehrman, "Misquoting Jesus: The Story Behind Who Changed the Bible and Why", [2005] states...

"What we have are copies many, many centuries later." [than the originals] "And these copies all differ from one another, in many thousands of places. As we will see later in this book, these copies differ from one another in so many places that we don't even know how many differences there are. It is easier to put it in comparative terms: There are more differences among our manuscripts than there are words in the New Testament."

And we don't really know how many differences there are because no one has ever counted them all that I know of. With the long history of scholars and laymen looking closely at the New Testament that fact is significant.

Christians look at their "Holy Bible" as if the words in their books have always been there exactly as they were written 2000 years ago, authored by the same people who have their names on them, and that it somehow answers the questions of our existence within those pages. It's a delusion to assume any of that. One that religion encourages in order to keep their sheep filling up their collection plates.

The Hebrew word found in the "Old Testament", "Sheol", was generally understood to mean death, destruction, and the grave; but was translated into the KJV with the old English word "hell." Christian translators in the 1600s and later were trying to translate the Latin translations of the Bible with the popular belief at the time of "going to heaven or hell." Before the English translations took place the word "hell' meant a hole in the ground and associated with mythologies similar to the Greek belief in an underworld, a world of the dead, called "Hades." The earliest translators thought that there was three Greek words in the New Testament texts that all described the mythological place being taught by their religion that was where the evil souls and the un-saved souls go for eternal "punishment." They were being taught by their religion that their "spirits", or "souls", were already immortal, continuing to exist after death.

The three different Greek words that were translated into the one word "hell" have completely different definitions according to the context of where they are found. None of those definitions were intended to be interpreted as a place where the dead go of Greek mythology. One of those definitions is the only place where our bodies go after death: to ashes, the grave, back to the dust of the earth from which we are made — of the atoms and subatomic particles that everything comes from. "Hades," the realm of the dead from Greek mythology, should have been written as the Aramaic word for the physical death and general word for destruction, "Sheol." Today's Abrahamic religions and the Greek writers of the New Testament have mistakenly interpreted Sheol to mean the same as that mythical hell of an underworld of the damned of the Gentile immortal soul beliefs of Greek mythology. That was not what the ancient Jews and early Jewish Christians believed at all. It was, however, what the translators of the New Testament believed because of their previous indoctrination into Greek mythology. It's only a "place" in the conventional sense like a graveyard. In the context of where it was placed it represents physical decay, decomposition, and destruction. The evidence will pan out. It isn't the Hades of Greek mythology because the Aramaic speaking Jesus and his disciples would not have believed in the Hades of Greek mythology. That means that the word "Hades" was written into the New Testament either by its original Greek writers or later Scribes, but would definitely not be something that Jesus and all first-century fundamentalist Jewish Teachers like him would have believed in.

Scholars know that ancient Jewish beliefs of life and death were different from the Greek and Roman mythologies. However, scholars don't write about that. To the earliest Christologies and Jewish beliefs before the Jewish War, you were either resurrected someday or you ceased to exist. As explained earlier, we are mortal souls who sleep in the dust awaiting a resurrection to Judgment where we are offered the tree of life or eternal nonexistence in Gehenna, the fire that will not be "quenched" (put out before burning everything up). And even the belief of a resurrection itself did not enter mainstream Judaism until about 2 centuries before Jesus. Before that point, Judaism was unclear on what death means but did not believe in an afterlife. As in all of human history, they were stumbling around in the dark looking for answers that couldn't be found.

The result of every individual deemed "incorrigible" will someday come to the final end of their existence, the "Second Death", the "unquenchable fire", the "hell fire", in Greek, the word, "Gehenna." Gehenna was translated into that same hell that Hades was translated into. We can't know exactly what is meant as the Gehenna second death mythology until that time comes.

And the third Greek word from the New Testament documents translated into the English word "hell" describes the place, condition, or dimension of the mythological fallen angels called, "demons," the "angels that sinned." Neither our bodies, "souls," or "spirits" are able to enter that place at all. It has nothing to do with us and yet religion has interpreted it to mean the same mythical "hell" of an underworld without any solid evidence that it means that. There is only one place in the Bible where this word appeared, also being deceptively translated into the word, "hell". That Greek word is "Tartarus."

When the word "hell" was placed over these three different Greek words in the New Testament documents being translated into English it's easy to see how the description of "hell" was perceived as a place under the ground ("Hades") of eternal (immortal soul doctrine) torment in flames

("Gehenna") where Satan and his demons are ("Tartarus"). But that's not what Jesus, his disciples, the Apostle Paul, and any other devout first-century Jews and Christian Jews would have believed at all.

1. "Tartarus"

The only place in the Bible where this word appears is...

2 PETER 2:4, (old KJV) "For if God spared not the angels that sinned, but cast them down to hell, and delivered them into chains of darkness, to be reserved unto judgment..."

Then there's this modern version...

(NIV, New International Version) "For if God did not spare angels when they sinned, but sent them to hell(b), putting them into gloomy dungeons to be held for judgment...) {b Greek "Tartarus"}

And then there's this modern version...

(RSV, Revised Standard Version) "but cast them into hell(f) and committed them to pits of nether gloom." { f Greek "Tartarus"} A clue to its definition is also found in...

JUDE 6, (NIV) "And the angels who did not keep their positions of authority but abandoned their home — these he has kept in darkness, bound with everlasting chains for judgment on the great day."

("The Jerusalem Bible: The New Testament" (Double Day 1968) "Next let me remind you of the angels who had supreme authority but did not keep it and left their appointed sphere; he has kept them down in the dark, in spiritual chains, to be judged on the great day."

Clearly this word was used to describe some kind of condition of restraint for the "angels that sinned" that are "kept in darkness", "in gloomy dungeons", and "in chains of darkness". However, in these statements themselves, or in the context where it is used, there is no implication that human beings, or "souls", or human "spirits" are in any way going to that same place or condition of Tartarus. Furthermore, there are no such things as angels or evil spirits. It was all made up as human imaginations evolved.

2. "Hades"

MATTHEW, 11:23, (old King James Version, KJV, emphasis mine) "And thou, Capernaum, which art exalted unto heaven, shalt be brought down to hell: for if the mighty works, which have been done in thee, had been done in Sodom, it would have remained to this day."

(New King James Version) "will be brought down to Hades..."

Notice that the writer here was referring to a city, "Sodom", which was brought down to "Hades". It was destroyed and brought down to ash and dust. It would otherwise "have remained to this day." It's not under the earth in a place called hell, it's in the dust of the earth.

MATTHEW 16:18, "And I also say to you that you are Peter, and on this rock, I will build my church, and the gates of Hades shall not prevail against it." The name "Peter" in Greek is a word that means rock, "Petra." Also, often translated "gates of hell."

The phrase "gates of Hades" represents the gates, entrance, of a graveyard. Peter and the "church" were expected to prevail against it by the prophesied soon coming "first" resurrection to eternal life after the soon expected "second coming" of Jesus. That, of course, never happened. So, it means that Peter did not prevail against the gates of the graveyard. He died like everyone else does. To point out a very important note about the above passage, Peter's real name was Simon, a very Jewish name of that era.

Here you see a Jewish Jesus changing a Jewish Simon to a Greek name like "Peter." It isn't something that the historical Jesus would have done. It was part of the agenda of the writers of the Gospels to change anything and everything Jewish about the Jesus movement for several reasons already discussed, like the Jewish war for independence from Roman occupation. To put it bluntly and correctly, the writers of the New Testament Epistles, Gospels, and Paul himself, were not on the side of the Jewish freedom fighters or Jesus's followers during that war with the Roman Empire. They were on the side of the Romans. That is the main issue, reason, and purpose behind the antisemitic leanings of the whole New

Testament. Some will say that it is just my opinion, but it is the most obvious one I could have. The Virgin Birth fabrication was completely inspired by opposition to the Jewish Christian belief that Jesus was a real, fully human, human being who was systematically adopted by God.

ACTS 2:26-27, 31-32,"Therefore my heart rejoiced, and my tongue was glad; Moreover, my flesh also will rest in hope. For You [God] will not leave my soul in Hades, Nor will You allow Your Holy One [Jesus] see corruption... He, foreseeing this, spoke concerning the resurrection of the Christ. That His soul [body] was not left in Hades, nor did His flesh see corruption [decay]. This Jesus God has raised up."

This was not intended to be translated into Hades except that it was Greek people who wrote it that way. Again, the Jewish Jesus, James, and the disciples and followers of the Way sect of Judaism would not have believed in a Hades of Greek mythology. To them it would and should have been written as Sheol. That is the Aramaic/Hebrew word to describe the physical death of destruction and the decay of decomposition. Which is exactly what it means to say that Jesus was not left in Hades, nor did His flesh see corruption.

3. "Gehenna"

MATTHEW 5:22, "But I say to you that whoever is angry with his brother without a cause [some documents omit "without a cause"] shall be in danger of the judgment. And whoever says to his brother, 'Ra'ca!' [in Aramaic "empty headed] shall be in danger of the council. But whoever says, 'You fool!' shall be in danger of (4)hell fire." (4 Greek = "Gehenna")

MATTHEW 10:28, "And do not fear those who kill the body but cannot kill the soul [alleged human spirit]. But rather fear Him who is able to destroy both soul and body in (1)hell." (1 hell, Greek = "Gehenna") Notice the word "destroy."

MATTHEW 18:9, "And if your eye causes you to sin, pluck it out and cast it from you. It is better for you to enter into life with one eye, rather than having two eyes, to be cast into 1hell fire." (1hell, Greek = "Gehenna")

A "King James Bible Dictionary" definition of the word "Gehenna" is, "the valley of the son of Hinnon (Joshua 15:8), near Jerusalem, where some Hebrew monarchs made human sacrifices to the god Molech (2 Chronicles 28:3). To discourage consideration of it as sacred place, the people were persuaded to dump their garbage there: this practice developed into the burning of fires in the valley to destroy the fly-invested filth: because of its abhorrent appearance it became a symbol for the fire of hell where the wicked are to be punished after the Last Judgment."

It, "Gehenna," is the Greek form of the Hebrew "GeHinnom", for the "valley of Hinnon" but was obscured into being the "Hell" of eternal torment when it was intended to mean an eternal "punishment" of non-existence, a Second (and final) Death.

MARK 9:42-44, "But whoever causes one of these little ones who believe in Me to stumble, it would be better for him if a millstone were hung around his neck and he were thrown into the sea. If your hand causes you to sin, cut it off. It is better for you to enter into life maimed, rather than having two hands, to go to (2)hell, into the fire that shall never be quenched — where 'Their worm does not die, And the fire is not quenched." [2 "Gehenna"]

What does that even mean, "Their worm does not die" and "the fire is not quenched"?

FIRST: In any fly-invested pile of trash, what kind of "worm" could there be in it that does not die? A maggot is a worm that becomes a fly. In that sense, the worm does not die. To say that the "worm does not die" could very well be an odd reference to the maggots in the trash at the "Valley of Hinnom" dump!

SECOND: How do you "quench" a fire? Of course, you put it out before it can consume all that it is burning. That entire last verse, 44, "where 'Their worm does not die..." was omitted by some documents of the text.

That last verse is a quote from a verse in the "Old Testament", "Book", called, "Isaiah." An Old Test book that the Gospel's appeared to have Jesus quoting from the most. Especially in the book called "Matthew," which was structured in such a way to appeal more to Christian Jews (Other than Matthew's version of the Virgin Birth tale) than the Greeks, Romans, or Egyptians. The author of Matthew was trying to show Jesus fulfilling prophesies and so he actually quoted from the Old

Testament for that purpose. We can be fairly certain that a historical Jesus did not say that. That realization just adds to how embellished fabrications are part of the New Test.

ISAIAH 66:23-24, “And it shall come to pass that from New Moon to another [months on the “Sacred Calendar”], and from one Sabbath [Saturday] to another, all flesh shall come to worship Me,’ says the LORD. And they shall go forth and look upon the corpses of the men who transgressed against Me. For their worm does not die, And their fire is not quenched. They shall be an abhorrence to all flesh.”

That is the last verse in the “Book” of “Isaiah.” Notice, “look upon the corpses” being burned in a trash-fire. What would you see?

Those Scribes who included that phrase in their copies of the writings could have actually believed that Jesus said it and understood that “Gehenna” is a representation of the trash piles burning in the valley of “Hinnom.” As well as the reference to those final verses in Isaiah it represents the “fire” that will consume all that it can consume. To “quench” a fire one puts the fire out somehow before it can consume all that is available for it to burn. So, an “unquenchable” fire is one that will burn everything that is somehow established for it to burn. The complication arises for those who have been taught the fantasy of the immortal soul and the hell of eternal torment. A fire that is not “quenched” will burn everything to ashes and dust. Right? It continues to burn all combustible material until nothing is left.

EZEKIEL 18:4,20, “Behold, all souls are mine; as the soul of the father, so also the soul of the son is mine: the soul that sins shall die... the soul that sins shall die.

LUKE 12:4-5, (Jesus supposedly said) “And I say to you, My friends, do not be afraid of those who kill the body, and after that have no more that they can do. But I will show you whom you should fear; Fear Him who, after He has killed, has the power to cast into hell [Gehenna]; yes, I say to you, fear Him.”

However, the tree of life is the logical process to any kind of indefinite longevity. Planting a seed and nourishing it with a benevolence higher than any mortal/physical being is expected to rise to might be a logical purpose of living.

One of the last books written and included in the “Old Testament”, or “Hebrew Bible” is

MALACHI 4:1-3, “For behold, the day comes, burning like an oven, when all the arrogant and all evildoers will be stubble; the day that comes shall burn them up, says the LORD of hosts, so that it will leave them neither root nor branch. But for you who fear my name the sun of righteousness shall rise, with healing in its wings. You shall go forth leaping like calves from the stall. And you shall tread down the wicked, for they will be ashes under the soles of your feet, on the day when I act, says the LORD of hosts.”

According to the actual biblical description of “hell,” our first physical death is a prelude to a resurrection. However, the human “spirit” itself does not experience the passage of time between that physical death and the resurrection. We need a living and "breathing" body for that. That requires the elements of the universe that made us, in this noticeably short lifetime, to provide the time necessary to teach ourselves the benevolence of empathy, the omnipotence of knowledge, and the patience of an eternal entity.

The time is always now. The past no longer exists and tomorrow never comes. Today's tomorrow is tomorrow's today. But we only live within the now. We don't exist outside of that unless we can evolve to a much higher life-form. A life-form which is not bound by the four dimensions.

EZEKIEL (Old Testament) *37:5-6,12-14*, "Thus says the LORD GOD to these bones: Behold, I will cause breath to enter you, and you shall live. And I will lay sinews upon you, and cover you with skin, and put breath in you, and you shall live; and you shall know that I am the LORD... Therefore prophecy, and say to them, Thus says the LORD God: Behold, I will open your graves, and raise you from your graves."

JAMES 4:12, "There is one lawgiver [some early documents add, "and judge"] who is able to save and to destroy." (Not written by the same "James the brother of Jesus")

HEBREWS 9:27,"It is appointed for men to die once, and after that comes judgment."

JOHN 3:16 (NKJV), "For God so loved the world that He gave His only begotten Son, that whoever believes in Him should not perish but have everlasting life."

As we have seen, the "soul that sins shall die". To "perish" in flames means exactly what it sounds like, everlasting death, non-existence. As it says here, it is the exact opposite of "everlasting life." It says, "should not perish but have everlasting life." The intended meaning of "perish" is the opposite of "everlasting life." The true interpretation is right there for you to see as long as you don't try to read into it what you've been taught to believe. It is the "fire" that is not "quenched."

PSALM 146:4, "When his breath departs, he returns to his earth; on that very day his plans perish."

(New Testament, the oldest letter by Apostle Paul) 1 Thessalonians 4:13, "But we would not have you ignorant, brethren, concerning those who are asleep [dead], that you may not grieve as others..."

This biblical and Jewish based theory proposes that Jesus was impregnated with the virtual "seed" of the "tree of life," which would sprout into a new-born of "God," "first-born," and "first of many brethren," after the seed itself died in the same sense as any seed must decay to produce a new life. It is the reason so many of Jesus's parables involve seeds and trees.

ROMANS 8:29, (NKJV, written by Paul himself) "For those God foreknew he also predestined to be conformed to the image of his Son, that he might be the firstborn among many brothers and sisters."

The Christianity that many of us was raised in made a hell of a mess for us to figure out. There was; Holy Ghosts, the hell of the damned, realm of the underworld, the wine and the wafer, church on Sin-day, hymns, and carols— bowing to a large plastic cross; standing, singing, then kneeling, then praying, then sitting, then kneeling, then standing, then singing, then praying again, to ad nauseam. We were taught to believe in going to some mysterious "heaven," a paradise in the sky somewhere, with our immortal souls, because we had water sprinkled over our heads when we were babies. They called that a "Baptism." It isn't correct. A true baptism is a complete immersion into water. The delusion that it creates plays a huge role in why so few people face the reality of our human ignorance in search of real answers.

REVELATION 20:14-15 (NIV),"Then death and Hades was thrown into the lake of fire. The lake of is the second death. Anyone whose name was not found written in the book of life was thrown into the lake of fire."

We are not immortal souls because the soul or "spirit in man" is not immortal. The BIG LIE continued into contemporary religions from non-Jewish sources until sometime after the Roman Empire made "Orthodox" Christianity their official religion. This mutated the early Christianity from the Jewish version of belief in a resurrection to the pagan belief in the "immortal soul doctrine," the mythology-based religions and thereafter mutating into the Jewish faith followed by Islam — that's all three of those Abrahamic religions that are based loosely on the early ancestors of the patriarch Abraham, which were stories told around campfires and retold for millennia.

The Christian Bible's "soul" is supposed to be the "living" breathing body merged with a temporary form of human "spirit" which "sleeps" in the dust of the earth while unconsciously awaiting a "resurrection" in which our bodies and spirits can be reassembled out of our original subatomic particles to be resurrected and then we can either participate in the overall purpose of the universe's existence or end up being incinerated in a Gehenna fire (hell) and perish forever, after given ample time and opportunity to avoid that — as any benevolent entity of an advanced life-form like the Designer of the universe would do. In other words, if we don't fulfill our purpose, "in dying we shall die." That is why it says in that very famous verse, "should not perish."

To "perish" does not mean just the physical death here. It is the action of being immediately burned up into permanent non-existence after a "Judgment" of some kind. The sentence explains itself with the opposite of "everlasting life" being everlasting death in the context of what it says. That is what the verse is saying. Most people have been reading something into it that isn't there because of the immortal soul doctrine and perpetual mistakes in interpretations that goes all the way back to the earliest days of Catholicism.

46

The Four Heavens

There are at least four different definitions of the word "heaven" used in the Bible. All four can mean completely different things or a combination of two or three of the four when spelled in the plural form. In the old King James Version (1611), the plural form was not used. This led to very confusing passages. The word "heavens" in the plural form is a combination but usually means everything that they could see above their heads. The ancient theist scribes wrote parts of the Old Testament and made copies of it from 2-3 thousand years ago living at a time when they didn't know much of anything. When used in the singular form in older versions of the Bible (KJV), "HEAVEN" is one meaning or plural, a combination of the different meanings. The word "heaven" in the older translations can mean clouds, stars, sky, birds, the sun, the moon, the planets in our solar system, wherever their gods are, or where the one God is, or combinations of that all in one word. Anything that was above their heads was in heaven.

We will take an in-depth look at this, along with other terms and phrases from the Bible. This is not research that is based on any denomination or institution of religion. This agnostic search involved buying and studying books on religion and The Bible over the course of many, many years from a wide variety of writers, scholars, intellectuals, and the prolific author, speaker, and teacher, a Professor of Religious Studies from Chapel Hill, UNC, Bart D. Ehrman. I looked at the views of theists, atheists, and agnostic writers and the many variations that there are.

The Bibles used here will be the New King James Version, the Revised Standard Version, and others; with underlining added for emphasis and information and comments added between brackets.

Critical interpretations and explanations unfolded by investigation through the processes of studying the evidence in dozens of books written by mostly agnostic religious scholars as well as every range of scholars from true believers to atheists. The Bible itself is the main source through this investigation.

It has been the widespread belief of biblical based religion that one of the definitions of "heaven" is a place where the "blessed", "redeemed", or "saved" will ultimately live in joy and peace as their final "reward." However, that definition comes from ancient mythologies that are not based on the original Christian or Jewish Bible.

After being conquered and occupied by one conquering superpower after another for millennia, the Jewish people were anxious to have someone, a human being or even the LORD God himself, to bring Israel back to self-rule. They didn't expect their God to come down Himself and take over in any form. That wasn't part of their belief system and history. It was always expected to be a fully human prophet with some of God's powers (healing miracles and such) and sometimes received in the ancient past according to their mostly fabricated and very embellished history.

The Teacher Yeshua had claimed to be a direct descendant of King David and Solomon, making Jesus the alleged rightful "King of the Jews."

Humans, and all other life, are temporary beings. A temporary being cannot be an eternal one at the same time. So, they were only expecting their messiah to be a fully human, human being to somehow confront and push the Roman Empire out of Judea to set up the government of their God with another Jewish King. He wasn't there to save the world and the New Testament Gospels are not trustworthy accounts in any way. They were not meant to be historically accurate in any way. The writers were making mythological fiction, just like their ancestors had done with the Greek and Roman gods and goddesses of their religious mythologies. The Jesus of the Gospels was like the Greek gods, born from a human women by a male god with magical powers.

The definition of "Messiah," which translated into Greek from the Hebrew/Aramaic is "Christ." {Wikipedia) {1} "the promised deliverer of the Jewish nation prophesied in the Hebrew Bible." {2} "a leader or savior of a particular group or cause." And (Britannica), "from the Hebrew "mashiah," "the anointed one," "anointed" has never meant divine."

In Judaism, they expected a king of the Davidic line to deliver Israel from foreign bondage and restore the glories of its golden-age of self-rule with their imaginary "God" in charge — which was really just a king and alleged "prophet" who were making the rules based on their sacred writings from the past and their primitive authoritarian traditions.

The other versions of Christianity, being led mostly by those who followed Apostle Paul's writings, was not a threat to Roman authority or the occupation of Judea and was exceedingly popular among Roman and Greek civilians. The small apocalyptic, messianic, Jewish sect of Jesus that started the Jesus movement was bound to fail and be replaced by pagan mythology. That's very much what happened. That very first Christology was an apocalyptic messianic Jewish sect who had to flee Jerusalem when the Roman army marched into Judea. The Jesus movement followers were dispersed and scattered over time, only showing up in obscure writings over a century later of Orthodox Christian priests. What little they recorded is hardly more reliable than the Gospels and for the same reasons.

The First Definition of Heaven

God's throne, or his present abode. Throughout the history of The Bible a "heaven" can be defined as the dwelling place of "Elohim" {"In the beginning, God"; the Hebrew word translated to English as "God" here is "Elohim"}. This is the first definition. "Elohim" is a plural name or word meaning "Mighty Ones," as the Hebrew word "Eloah" means "Mighty One".

This could be described as a family name; the Creator Family, the Designer Family, the God Family. It could also describe the name of a species of life. It probably just means the plural form of gods that they thought existed.

GENESIS 1:26, "Let us make man in our image, after our likeness."

"God" is actually a plural name but used predominantly in religion as a singular term. The Hebrew Bible phrase, "the LORD God" is almost like a first and last name. This Old Testament word "LORD," spelled thusly, has been translated from four Hebrew letters of their alphabet which has been translated into the English letters "YHWH." These letters are sometimes translated as "Yahweh," which is simply putting vowels in for pronunciation. The word "Jehovah" comes from languages in which the "Y" sounds like a "J," and the "W" sounds like a "V"; with vowels inserted. In comparison, the Hebrew name "Yacob" becomes "Jacob." So, "Yahweh" can become "Jehovah."

In ancient Israel, the "YHWH" name was not spoken aloud except by the High Priest once a year. It is the sacred name of God, a kind of spokesperson of God, but not a name that observant Jews are allowed to utter to this day. As stated above, it was a big part of why Matthew frequently used the term "Kingdom of heaven" instead of "kingdom of God." He felt that using the name of God in this way might anger some fundamental Jews.

It is obviously apparent that God made it a commandment not to "use" His "name" in "vain" and ordered the Israelis before those commandments were given to never speak His name out loud. There seems to be a firm desire from this God to not be acknowledged by means of a name in any way by humankind, at first. As if He knew that associating any name with "Him" would lead humankind to misrepresent "Him" all through the ages of man's misrule over the earth. This "God" did not want us to give "Him" a name or misuse the name that "He" told Moses.

When the LORD appeared to Moses from the "burning bush" in...

EXODUS 3:13-15, "Moses said to God, "If I come to the people of Israel and say to them, "The God of your fathers has sent me to you," and they ask me, "What is his name?" What shall I say to them?" God said to Moses, "I AM."

This is also translated as "I AM WHO I AM," and "I AM WHAT I AM", and "I WILL BE WHAT I WILL BE". A note in the Concordance of the Revised Standard Version reads, "The word "LORD" when spelled in capital letters with the last three letters smaller, stands for the divine name, "YHWH" which is here connected to the verb "hayah, to be."

"God also said to Moses, "Say to the people of Israel, "The LORD, the God of your fathers, the God of Abraham, the God of Isaac, and the God of Jacob, has sent me to you"; This is my name forever, and thus I am to be remembered throughout all your generations."

Until that moment the Israelis did not know the name of their God. It was generally believed that their God was too sacred, powerful, and unknown to be given a name. In ancient Egypt there were many gods, all of which had names. In this way they could explain the powers and forces of nature by attributing different gods with such things as; the earth, the sea, and the sun, and so forth.

If this "LORD God" thing was the true Designer, He would not want to be manipulated in that fashion or be associated with fictitious gods. However, the Bible does not say that there are no other gods, just that "You shall have [worship, acknowledge] no other gods before {besides} me." It was very unusual for a god in that era to be a god of everything, or "monotheistic" {the belief in just one God}, and without even a name. By knowing your god's name, you could claim to speak for him, misrepresenting his wishes to the masses. As we can easily understand now, considering religious history and the history of the world so far, a Designer would not want anyone speaking for him or representing His wishes, or His will. And as stated above, the true Designer of the universe is Impersonal. IT has no desire to get involved with human while we are still evolving from animals. Especially when we were in the primitive stage of development. It would not make sense.

If any God or intelligent designer really exists, "He" simply would not want anyone to speak for Him until He is ready to speak for Himself if He feels it necessary.

"He" is not communicating with us at all. Certainly not in any book written by any human being. Is "He" speaking to us now? Of course not! "He" doesn't even exist to us right now. Being from different dimensional planes, we don't really exist to Him, similarly to how He doesn't really exist to us. Time, of which we experience in our present space-time universe, does not exist to a Designer in the same sense that it does to us. IT is residing in a different dimension (for lack of a better term) which is not within our space-time bubble. If we don't destroy our species and evolve to a higher form of existence, we will also be able to occupy the dimension of the Designer as part of the overall goal, the underlying purpose, of the universe.

The first definition of the word "heaven" in the Bible is the place, or existence, or dimension, in which this "Father" and allegedly his "firstborn Son" presently reside. Their throne is in "heaven" as in the "Lord's Prayer," *MATTHEW 6:9*, (KJV) "Our Father, who art in heaven." The book of *JOB 22:12* states that God is "high in the heavens." Which could be somewhere beyond or among the stars as *PSALM 14:2* states "The LORD looks down from heaven." According to *MATTHEW 11:25* the Father is "Lord of heaven." *MARK 12:15* says that God's angels are "in heaven."

We have no secret wisdom inside of us at our disposal to adequately envision the residence of a possible "Supreme Being" or even pretend to know what "IT" is who/what designed and set in motion this vast and complex universe. My guess is that the MIND behind the design of the universe is from a higher dimensional plane of existence. IT would not even be able to exist in this universe if IT is made completely of pure energy, not even the kind of energy that our universe can create from itself.

So far, we have established that at least one definition of the word "heaven" includes the present residence of the biblical "God." Wherever it is, whatever it is, "IT IS." I cannot read any more than that into it and no one else should. Except for one more thing. The God of the Holy Bible and any religion is not the Impersonal Universal Consciousness that would be the real Designer of the universe.

The Second and Third Definitions

The second definition is the universe outside of the earth's atmosphere is the "stars of heaven" which are "high in the heavens." The third definition is the atmosphere above us itself, literally in the clouds, from one horizon to the other. The early translators did not distinguish specifically between where the clouds are, where the stars are, and where God's throne is. If it was above their heads, so they called it "heaven." A good example of this is found in *MATTHEW 24:29-31*, while Jesus is speaking about his second coming...

(RSV) "Immediately after the tribulation of those days the sun will be darkened, and the moon will not give its light, and the stars will fall from heaven, and the powers of the heavens will be shaken; then will appear the sign of the Son of Man in heaven, and then all of the tribes of the earth will mourn, and they will see the Son of Man coming on the clouds of heaven with power and great glory; and he will send out his angels with a loud trumpet call, and they will gather his elect from the four winds, from one end of heaven to the other."

A more accurate translation at the end of that last verse would be, "from one end of the horizon to the other." Scholars call the Matthew 24 prophecy the "Olivet Prophecy" because Jesus and his disciples were speaking privately on the Mount of Olives. The Gospels of Mark and Luke have their own versions of this event because it was originally in Mark before Matthew and Luke borrowed it for their versions of the story. A book in the New Testament that scholars are sure was written by Apostle Paul is called

1 THESSALONIANS, the oldest book in the NT and written by Paul himself, clears up what Matthew 24:31 meant by the statement "from one end of heaven to the other." NOTE; that I have touched on this passage in previous chapters. These next verses are used as doctrine involving the recent ignorant evangelical fad called "The Rapture."

1 THESSALONIANS 4:15-17 "For this we declare to you by the word of the Lord, that we who are alive, who are left [are not dead] until the coming of the Lord, shall not precede those who have fallen asleep [have died]. For the Lord himself will descend from heaven with a cry of command, with the archangel's call, and with the sound of the trumpet of God.

And the dead in Christ will rise first; Then we who are alive, who are left, shall be caught up together with them in the clouds to meet the Lord in the air..."

So, to gather people from around the world, from the four winds, from one end of the planet to the other, to converge on Jerusalem as newborn children of God; they allegedly, supposedly, meet together in the upper atmosphere according to Paul. When you look at the sky, it arches from one horizon to the other. This is referred to in the Bible as the "firmament." What Jesus meant in saying, "from one end of heaven to the other" is actually from one horizon to the other. The "four winds" are the four directions from Jerusalem: East, West, North and South.

GENESIS 1:20, "the fowl that may fly above the earth in the open firmament [arch of the sky] of heaven."

It has been assumed that Elijah was taken up to "heaven," to God's throne. That is just a mistake.

JOHN 3:13, {RSV} "No one has ascended into heaven but he who descended from heaven, the Son of Man." This Gospel is the most unreliable because of its late writing at least 70 years after Jesus.

Popular religion has been trying to explain away this verse because of their beliefs of "going to heaven or hell." He would have been talking about being physically carried into heaven. However, some Old Testament prophets were also "taken" up into "heaven" before they physically died.

2 KINGS 2:11, "And as they still went on and talked, behold, a chariot of fire and horses of fire separated the two of them [Elijah and his son Elisha]. And Elijah went up by a whirlwind into heaven."

As you will see, this is not the "heaven" of widespread belief or of God's throne. Elijah expected it. Just before it happened, "Elijah said to Elisha, 'Ask what I shall do for you, before I am taken from you.'" In that era, the citizens of Israel and Judah had insisted on having "kings" to rule over them. They had appointed "priests" of the "LORD" to give "council" to the kings. The King of Israel at that time had disobeyed and ignored the council of Elijah {the Priest of God}. Elijah then told the king that he, the king, would die.

2 KINGS 1:17, "So he died according to the word of the LORD which Elijah had spoken. Jehoram, his [the king's] brother, became the king in his stead."

So, Elijah had been the priest of the LORD to the now dead king. Being the son of Elijah, Elisha was chosen to become the new priest to the new king, Jehoram.

After Elijah was "taken up"...

2 KINGS 2:13-14, "And he [Elisha] accepted the mantle of Elijah that had fallen from him and went back and stood on the bank of the Jordan [River]. Then he took the mantle of Elijah and struck the water saying, "Where is the LORD, the God of Elijah?" And when he had struck the water, the water parted to one side and to the other; and Elisha walked across."

Elisha had just become the new priest to the new king. God surmised, logically, that if Elisha would serve Him to the new king effectively, Elijah's presence might somehow hinder Elisha's authority, maybe even undermine it, and therefore impede his ability to serve his god. However, would this God end Elijah's life for that purpose? Wouldn't this premature trip to the "heaven" of popular tradition mean that his life as a material flesh and blood human on Earth was over? According to some scholars, about ten years after this event takes place, Elijah sends a letter to Jehoram from somewhere.

And of course, all this is according to the texts and proper interpretation based on how The Bible was written. I am not saying that any of this happened.

2 CHRONICLES 21:12, "And a letter came to him from Elijah the prophet..."

Did Elijah write a letter and send it from "heaven"? Of course not, because Elijah was "taken up" to the clouds, or at least high enough out of sight, and brought back down to someplace else on earth to live out the rest of his life. It was not the "heaven" of God's throne, or of popular myth, but of the "heaven" of the clouds.

The prophet Enoch was also "taken" to someplace in the same manner. When I found another tale of someone being "taken" and it looks like a remarkably similar situation, I felt confident that I had the right interpretation, particularly when so many other of my interpretations lined up with corroborating evidence and logic. Some might picture a UFO picking up someone from the ground and into their craft to be brought back down somewhere else to live out the rest of his life without the threat of violence against him.

GENESIS 5:22,24, "Enoch walked with God... for God took him."

HEBREWS 11:5, "(NKJV) By faith Enoch was taken away so that he did not see death, and was not found, because God had taken him; for before he was taken, he had this testimony, that he pleased God."

Being taken doesn't mean taken to the heaven of popular mythology. Notice that "he did not see death and was not found." (unknown author, not Paul)

Although the Greek translated to English word "taken" can be appropriate, it is translated differently in...

COLOSSIANS 1:12-13, "(RSV) giving thanks to the Father, who has qualified us to share in the inheritance of the saints in light. He has delivered [saved] us from the dominion of darkness and transferred us to the kingdom of his beloved Son, in whom we have redemption, the forgiveness of sins." (unknown author, not Paul)

After reading that you might say that it is evidence of the going to heaven or hell doctrine. However, the acclaimed scholar Bart D. Ehrman says, "Did Paul write Colossians? According to most scholars no — Paul did not write Colossians." (*erhmanblog.org*)

If you wanted your beliefs about Christianity to be taken seriously you could write a "letter" to a "church" or someone and claim that Paul wrote it himself or Peter or James or John or any of the characters in the Gospels, even Judas. There were many Gospels and stories written during the first centuries in which the vast majority were not included in the New Testament and much of it was even destroyed by catholic leaders, priests, and heresy hunters whenever it contradicted the official doctrine of the "Orthodoxy" which became the "Roman Catholic church." After the Roman empire fell, they simply changed their name to "Catholic Church" and for several thousand years they eventually ruled most of the known world. The parts of the world that was well known at the time among the Romans and Greeks.

The NKJV has "and conveyed us" instead. It is believed that Enoch was in danger of being killed by the people for speaking out against them. God had "transferred", "conveyed", him to another place "so that he did not see death," to be killed by a pious and angry mob.

ACTS 2:29,34, "(NKJV) Men and brethren, let me speak freely to you of the patriarch David, [who lived centuries before them] that he is dead and buried, and his tomb is with us to this day... For David did not ascend into the heavens..."

ACTS 3:20-21, "and that He may send Jesus Christ, who was preached to you before, whom heaven [God's throne] must receive until the times of restoration of all things, which God has spoken by the mouth of all His holy prophets since the world began."

The New Testament, the book that Christians believe is the very word of God, says that everyone who is dead, except Jesus, is "asleep," not in any kind of heaven.

The first meaning reported is the throne of God (maybe the Designer) somewhere beyond the stars or a higher dimension. The Designer would not logically exist in, or be composed of, his own creation, this advanced holographic program. He is clearly reported to be somewhere else. Any interpretations to the contrary are probably false. He is not in the birds, flowers, bees, or trees — they are all parts of this "living organism" (Hawking/Hertog) that is the universe which is an advanced holographic projection. Although He is in the "tree of life," the tree of life is not here. It has not been replanted yet. "He" is not in us! The Bible says that "God" is in "heaven." Those who claim that God is inside us are mistaken. Our "spirit" is a strictly human form.

The second and third meanings clearly does not distinguish between the clouds and the stars, labeling both as "heaven." Ancient peoples did not understand how separate clouds and stars really are. The clouds clearly were below the stars, but it was all part of the "firmament" of "heaven" to them. It was all above their heads. That was as far as they knew.

The Fourth Definition

The fourth possible meaning of the word "heaven" could have been represented as the "kingdom of heaven" by

"Matthew" alone! The rest of the New Testament records it as the "kingdom of God." The Old Testament/Hebrew Bible alluded to its practical alternative meaning in their use of the Hebrew word "Elohim", translated as "God", but describing a plural form, or family name. The "kingdom of God" could be the Family of God (Elohim), presently consisting of the "Father" and possibly a firstborn "Son." Keep in mind that all this is speculation based on sound judgment, intense research and remember that the Bible is just a book. If there is anything to learn from it, it must include a logical purpose for the existence of everything. A "kingdom" can represent a form, or family, of a species as well as a political system ruled by a family-line, or lineage.

The thirteenth chapter of Matthew has eight verses in which it quotes Jesus saying, "the kingdom of heaven." He did not mean it is "in" heaven, the writer would have written it that way if he had. The "kingdom" is "of heaven" in that it is owned and ruled by "God," or it is simply the species of "God" itself. A kind of God-phyla, a God family. As stated, Matthew's author was Greek, and part of his agenda was to link things that Jesus did with Hebrew prophecy. He knew that any Jewish readers of his book might be offended by the loosely used name of God and therefore wrote it as a kingdom of heaven in places where Jesus is quoted.

The early Jewish Christians considered the "Apostle Paul" to be a dangerous heretic. That turned out to be true. History's list of massacres and persecution of the Jewish race is predominantly and nearly exclusively Paul's fault. The Gospel writers who inflamed a hatred of "the Jews" from that first century on were influenced by Paul's "Christology." It is a well-known proven fact that this "Paul" {Saul}, the writer of some (7) of the New Testament "Epistles" and mistakenly attributed to writing most of the others, had spread a doctrine of licentiousness after a faked conversion. Before his alleged conversion, he had severely persecuted the first followers of Jesus by his own admission, most likely intimately involved in the "Stoning of Steven" as recorded in *ACTS 7:54- 8:1*.

The word "gospel" comes from the Greek language for "good news".

LUKE 4:43, "I must preach the good news of the kingdom of God."

MARK 1:14, "preaching the gospel of the kingdom of God."

The Gospel of John is the last "Gospel" written, as much as eighty years after Jesus died. It is curious to note that the writer may have been part of a community of Jews that had escaped the destruction of Jerusalem by living in Greece and was originally influenced by Jewish Christians and the Gnostics rather than the "Pauline" Gentile-Christians. The writer had been taught some facts about Jesus and the "Way" early on but became disassociated and disenchanted with non-Christian Jews when all Christians were barred from the synagogues. The Gnostic and Pauline Gentile-Christian teachings became the most popular. Why were "Christians" barred from entering synagogues around the late first century or early second century? Because some "Christians" were trying to sell their versions of Christologies to the Jews there, particularly the evangelical Pauline Christology where people were pushing their beliefs on others. (you know the type)

Again, Jewish Christians, like the original Jesus movement, were not evangelically seeking converts. Paul wrote about trying to convert Jews and others and it's all over the book of Acts. However, the original Jewish Sect of Jesus, both before and following his crucifixion, never did that. They were not trying to convert anyone. They were preparing themselves for the tree of life, the Way Jesus thought that he did in the benevolent nature of serving others. That was their primary doctrine. Their secondary concern was in keeping the Jewish laws and statutes in a benevolent nature. It was never about making disciples of the world. That is the most fabricated thing about the whole New Testament.

Which also brings up the obvious fact that if they were strictly an exclusive sect to a specific audience of Jews, it was never intended to even be a religion at all. It was in preparation for what they expected to happen soon. The "kingdom of God is at hand." That is also translated, "has drawn near" and probably in some translation, "is coming soon." That explains why the book of Acts makes it perfectly clear that Jesus never instructed them on how the Gentiles could be converted. The subject never came up while he was alive because it wasn't their Way.

The Gospel of John went through many changes before the last version was finished, which is all that we have in actual documentation. The writer refers to Jesus sometimes as, "Rabbi," "Lamb of God," and "Messiah", none of which refers to a divine personage. A "Rabbi" is clearly a man but must be over 40 years old, the "Lamb of God" is a sacrificial animal, and the "Messiah" is prophesied in ancient Judaism as being a man, not divine, chosen to bring God's rule back to their "Holy Land." Scholars think that the first 18 verses of John were added to the text, influenced by some of the Christology called "Gnostic" who believed, among many things, that Jesus was completely "God" in the flesh from before birth and the "Word" that existed "with God" from the beginning of Genesis. However...

JOHN 1:12, "But to all who received him, he gave power [also translated, "the right" and "the authority"] to become children of God; who were born, not of blood, or of the will of the flesh, nor of the will of man, but of God."

JOHN 3:3, "(NRSV) Very truly, I tell you, no one can see the kingdom of God without being born from above."

This is also translated as, "born again" and "born anew". The book of "Hebrews" was written also to some degree with influences of early Jewish Christians. Which describes them as...

HEBREWS 12:23, (NRSV) "to the assembly [translated mostly as "church"] of the firstborn who are enrolled [also translated, "registered"] in heaven."

When one is "enrolled" or "registered" in someplace it does not mean that person is already there, or necessarily be going there at all. According to this author these people were specifically chosen for a particular part of God's plan. To be supposedly “born again/from above” when “He” returns from “heaven.” While they were still alive, they were merely “enrolled” or “registered.” To use the term “firstborn” is to imply that Jesus was the first to be born literally “of God” and that there will be others thereafter at some point. You really can’t interpret it otherwise. Again, this next verse from “Romans” displays Paul’s doctrine on this issue.

ROMANS 8:29, [accepted by scholars as written by Paul himself] “For whom he foreknew, he also predestined to be conformed to the image of his son, that he might be the firstborn of many brethren.”

It may be that I repeat these most important passages in other chapters and for good reason. It should be understood that being the “firstborn” does not mean that other people will be born like the virgin birth story. Jesus was simply a man of whom it was believed was given a “Holy Spirit” of God at his baptism and supposedly obeyed his soon-to-be “father in heaven” through that 3-year ministry until he was crucified and was THEN born of God when God resurrected him according to everything we know about what the first Christians really believed about Jesus.

Christianity went from believing in a Messianic rule and apocalyptic judgment on Earth someday soon to an afterlife of going to a “heaven” or “hell” when decades went by without the second coming actually coming.

“God’s Problem”, by Bart D. Ehrman, pg. 255-256:

"The earliest Christians believed they were living "in the last days." Their Lord had himself been an apocalypticism who warned the people of Israel to repent before it was too late, for "the Kingdom of God is very near" (*Mark 1:15*). And Jesus had been a follower of John, who indicated that the "ax is already laid at the root of the tree"— in other words, that the apocalyptic judgment was soon to begin. Jesus' own followers thought that he would be the one to bring that judgment, that he had ascended to heaven but would soon return to judge the earth and bring in the Kingdom of God as the messiah. They expected it all to be imminent.

But the days of waiting turned into weeks, then into months, then into years, then into decades. And the end never came. What happens to a belief that is radically disconfirmed by the events of history?

What happened in this instance was that the followers of Jesus transformed his message. In some ways the apocalyptic hope can be understood as a kind of divine timeline in which all of history is divided into two periods, this wicked age controlled by the forces of evil and the coming age in which evil will be destroyed and God's people will rule supreme. When the end did not come as expected, some of Jesus' followers transformed this temporal dualism (this age versus the age to come) into a spatial dualism, between the world below and the world above. Or put differently, they shifted the horizontal dualism of apocalyptic

> expectation of life in this age versus life in the age to come (horizontal dualism because it all takes place on this plane, here on earth) into a vertical dualism that spoke instead of life in the lower world versus life in the world above (with an up and down). In other words, out of the ashes of failed apocalyptic expectation there arose the Christian doctrine of heaven and hell."

Maybe, a definition of heaven is a kingdom of another kind. A family, maybe a life-form, maybe even the family of the highest life-form. Maybe so, I don't know.

We can assume now that the historical Jesus, the real man himself, made it his goal to get his Jewish God (or whatever god) to adopt him into ITs own species of life through the tree of life. He understood the answer to the ontological dilemma of knowing what the purpose of the universe is and the process of procreating an entity like God. We must seriously consider what a real Designer of the universe might be trying to accomplish, but it would be way above the knowledge of those first century humans and barely within our own ability to understand it now. Since we cannot know what is ahead of us, we can, however, begin to look and think about things differently in light of this new knowledge. The universe is us, because we are the universe down to the smallest possible level. Every living thing comes from the same atoms that were born in stars. The world does not revolve around us, it revolves around a star, which revolves around a galaxy which revolves around a Black Hole along with maybe 2-trillion other galaxies in the known universe. If we can see the stark reality in this, we can know that its creator, if there is one, would be a benevolent entity of some kind. Not like any God that you or I can imagine other than some kind of benevolent and Impersonal Universal Consciousness.

Through the words and deeds of that "Jesus" from the Gospels, we can actually have a promising idea now of what we have to adopt as our own Way of thinking. The things that Jesus said should replace all of the rituals, sacred writ, Creeds, prayers, church literature, and holy doctrines.

If there is an overall purpose behind the universe by a MIND of some kind, the level of benevolence which Jesus may have tried to embody and teach is not likely just a coincidence. If you are willing to consider that there is a real Impersonal Universal Consciousness behind this designed design with an overall purpose directly involving sapient and sentient physical life within this biofriendly living universe, what else could it be?

47

Oral Traditions

"Oral Traditions" are stories that were passed on "orally" (by word of mouth) about all kinds of religious beliefs in antiquity. It was Oral Traditions that made up what became known as the Old Testament/Hebrew Bible and then 2000 years ago it was stories told about what Jesus did and said that ended up in the four "Gospels." They are called "Oral Traditions" because they were exchanged between people by those who heard them from word of mouth. Not necessarily from Jesus himself but from people still around who heard it from second-hand or third-hand accounts. As the decades went by, there became many storytellers repeating Jesus' alleged deeds and words getting around which finally ended up getting to the Greek authors of the Gospels who wrote them down between 70CE and 140CE.

There were many other similar stories about Jesus and the first Christology that was either silenced by heresy hunters in the years leading up to Catholicism (fourth century) or those stories were rejected by those who put together the New Testament. Mark was the oldest of the chosen Gospels to be included in the New Testament, written about 40 years after the crucifixion. Like the children's game "telephone," the stories went through changes as they got passed around, but those changes never made Jesus (Bart D. Ehrman) "look less interesting, less powerful, or less knowledgeable as the traditions got circulated." Some embellishments, additions, deletions, and other kinds of changes took place before and after they were included into the New Testament "Canon" when it was established around 400CE.

> Ehrman: "You can tell when something is written that is not part of the structure or message, not part of the composition that the person is composing at the time. It might be more poetic, or it interrupts the content of the composition."

Some of what Jesus did and said in those Gospels are certainly questionable to have come directly from Jesus, but it is all speculative in any event. Anyone who claims that the historical Jesus never existed were biased in their "research" from pro-atheist circles who only look for historical ways that could allow them to assume that he didn't exist at all. They demand that we prove that a real Jesus existed. However, the circumstantial evidence material available now makes it logical to say, if nothing else about Jesus, that he was a real human being. There is too many important factors to consider that would deny the existence of a "Yeshua" who's name became the transliteration of "Jesus," which was less Jewish and therefore more popular with the predominantly non-Jewish citizens of the Roman Empire.

Jesus was a fully human man who was a Jewish Teacher of an apocalyptic messianic sect and was crucified for subversive activity against the Roman occupation of Judea. The claim by devout atheists of Jesus never existing is called the "Mythical Jesus," not to be confused with the "Mystical Jesus" from some of the Gnostic Christologies of that first and second centuries who believed, in some cases, that Jesus was a very strange and obscure being; like a man being possessed by a god, a god posing as a man, a projection of a man by a god, and some other strange beliefs — including that Jesus was "with God" since the beginning of Creation.

Saying that the person Jesus/Yeshua never existed is one of those easy answers that usually doesn't turn out to be true when an open-minded and closer examination of the facts is conducted. There are too many people from the story that we know really existed. His brother James is one. The New Testament authors tried to hide James in their writings. He was too Jewish by reputation to be included in their antisemitic versions of the story. That is a fact that cannot be denied. The writers of the Gospels embellished their stories to make the Jewish population look responsible for the crucifixion. They weren't. It was a survival strategy to blame "the Jews" to prevent Roman authorities from going after the Greek gospel writers and those of their Christological group of believers.

It is interesting to note that the documents of the Christian New Testament are filled with scribal changes and mistakes but the documents of the Hebrew Bible (as far back as we can go) and the Quran (Islamic scripture) doesn't have nearly as many. This reflects that the scribes who preserved the Hebrew

Bible and the Quran were much more careful than those who worked on the New Testament writings. Knowing Christians as we do, they changed things in the texts that did not agree with their beliefs. In one extreme case, scribes added to the last portion of the Gospel of MARK well after it was written. They didn't like the original ending and that allowed them to insert their most delusional beliefs into it by putting words into Jesus' mouth. I'm not kidding. Scholars know this very well. Most modern translations of the Bible include at least a footnote on the fact that the last half of the last chapter of Mark was added to the text at some later date.

When stories were told and retold about Jesus there are three different basic kinds that appeared in the Gospels and other first century writings about Jesus. Either it appears to be {1} "something that someone made up on the spot," (Bart Ehrman) in a moment of imagined inspiration or it was {2} based on the actual words and deeds of Jesus that were witnessed and embellished to some degree as they circulated Christian groups {3}, some regional mythologies were being incorporated into the Jesus stories and {4}, there is a definite agenda on the part of the Gospel writers to portray doctrines that they favored and make the Jewish people look bad in the eyes of the Roman citizens during a time of war between Rome and the Jewish population. Antisemitism was not the primary reason for the vilification of the Jewish race, it was to lure local Roman and Greek mythological believers away from the Jewish Christology and their requirement of converts to essentially become Jewish in order to become a follower of this new hybrid religion of a Jewish Sect with a dramatic story.

The New Testament Gospels has three of the four being so similar that two Gospels, years and miles apart when written, used the first one as a source along with using other early writings, folk lore, and accounts to make their own personalized versions with similar agendas. Matthew and Luke are clearly their personal and/or tribal versions while using Mark as a loose guide. They both used the sayings called Q, or "the gospel of Q." "Q" was a book of sayings that are associated to have came from the mouth of Jesus by the earliest of followers in the Jesus movement not long after the crucifixion. We don't have all the sayings there could have been because all we have are snippets used in those "narrative gospels." There were many writings by various early Christologies, some called Gospels that were rejected by the very early 5th century Catholic Fathers who were deciding to include which books and letters to combine together to make the New Testament over 4 centuries after Jesus. Most of the ones that were rejected had strong indications that it was made up. Like the Gospel of Peter which had a giant walking and talking cross. However, the strongest indicator of what was rejected was when there was differences in doctrinal beliefs that the Orthodoxy, which was based the most on Apostle Paul's Christology, did not want included. Paul's was the easiest of them all to convert to and so, the most popular. It was virtually like giving out a license to break all the rules. Many people were attracted to that ideology. Many still are.

The book of Acts has many made-up speeches by the author being told by different people, but mostly by the main character; Apostle Paul. The best scholars have reported a similarity in every speech that indicates they were written by the same person, the author of Acts himself, Luke. However, even though we have known for a long time that the author of Luke and Acts are the same, they were not the same Luke who was Paul's second assistant, traveling companion, and secretary. In addition, most scholars have been able to date the Gospel of Luke to "some time in the second century." (Burton L. Mack) or late first century. Either way, the real Luke would have been at least 90 years old when the Gospel of Luke was written and he would have been over a hundred when Acts was written. Even though the character of Luke was the only educated person in the Gospel stories capable of writing one of the New Testament books, I don't believe he was still alive when Luke's gospel or Acts was written. The disciples were among the lower-class peasants of rural Galilee, fishermen like Peter, for example. These people could barely read, were Aramaic speaking Jews, and the New Testament books were all written in the standard educated Greek of their day. Furthermore, the earliest followers of Jesus were apocalyptic, messianic Jews awaiting and anticipating "the end of the world" (age) and the coming of a Messiah. The thought of recording the moments of Jesus' life leading up to his crucifixion and after, was undoubtedly not on the minds of anyone who was close to the situation. Any writings by the original followers of Jesus would have been eradicated years later by Orthodox "heresy hunters" because it didn't describe the doctrines of Paul's Christology that had taken over as the most popular one by the end of the Jewish War.

In fact, the heresy hunters were looking for people of their own time who were teaching anything that wasn't the accepted Orthodox Christology, even the original Jewish Jesus movement was considered heretical. Not enough of these facts have been investigated by not enough scholars as to the connection between the Jewishness of Jesus and why his true Christology faded into obscurity. Why not? Because that would require Christianity looking uncomfortably close at what the truth is. Those heresy hunters went around destroying documents written by the remnants of the original Jesus movement. As the popular Christology evolved toward becoming Orthodoxy, they effectively rewrote history by claiming that the followers of Jesus and "Apostles" had always been of the same doctrines of beliefs as the Orthodoxy did centuries later. That was when everything Jewish about Jesus and his first followers had mutated into a Greco-Roman mythology. The hunters firmly believed that the ways of the Jewish Christians had to be silenced. They had variable reasons for wanting this. The Gospels had both subtle and extreme differences but generally agree on the reasons for their embellished and fabricated versions of history.

> *Bart Ehrman*: "They were written at different times, different authors, even when they repeat the same stories, the authors will change the story in one way or another."

Or they were giving different versions of the same story while the basic outlines are the same. The "Virgin Birth" stories (fiction) are so obscure that Matthew's account varies greatly from Luke's other than the (Ehrman) "basic outline of being born in Bethlehem to Mary." Matthew's and Luke's authors made up the virgin birth story but the rest of their versions were composing the basics of what happened as others were telling it in oral stories, reshaping it with their own beliefs and a Greco-Roman mythological spin, including tales of miracles (that didn't happen) with antisemitism. Adding to that the fact that as the stories spread very quickly and were retold orally from one person to the next, there were already changes to them before the many writers got their hands on them. So, how faithful is the accuracy of the retelling? There are lots of logical evidence that many passages and the whole Virgin Birth story was made up.

What might they be trying to emphasize in their versions of Jesus' life and death? What happens when stories are told and retold for 45 years (Mark), or 65 years (Matthew), or 70 years (Luke), or 80 years (John) after the events that they are writing about. (Bart Ehrman), "decade after decade, year after year, people tell one story to the next person and then the next. And it goes to various parts of the world. They're telling it in different languages, and they're influenced by different things, different views, different assumptions."

Who in their right mind would think that these stories are the infallible and inspired words of a real God, or even be historically accurate, especially when there are so subtle and vast differences between all Christian writings? Couldn't a

God of everything get them all to say the same things? Couldn't an Impersonal Universal Consciousness get all of the writers of the New Testament to be interconnected like subatomic particles? Even the passage that says "All Scripture is inspired by God" is an absurd statement because the person writing that was lying about being Apostle Paul.

God had nothing to do with it. That is simply impossible, and to believe otherwise is just fooling yourself and everyone else. The truth is much more plausible and amazing but upsetting in the scope of deception by religion. Believing that divinely inspired spontaneous miracles are possible in our present reality is a deception, just like the deceptive verse that says,

MATTHEW, 19:26 (NIV) "Jesus looked at them and said, 'With man this is impossible, but with God all things are possible.'"

Which has been interpreted to say that God has supernatural powers to use for anyone that He wills to use it for or against. That is the same old ancient mythological belief system in gods. It has no validity in today's world. They didn't know any better 2000 years ago. We should know better now!

Very few quotations of Jesus in the Gospels were quoting what was overheard during his life that circulated by oral storytelling. That resulted in a quagmire of bits and pieces of things he most likely didn't say (he was a real man) and undoubtedly much of it got lost in translation and interpretations based on Greek or Roman mythologies. Much of it was even embellished and exaggerated versions, or just fabricated for religious propaganda with disputes between separate groups of opposing Christologies.

> *Bart Ehrman*: “Most New Testament scholars just make assumptions about it, about the oral traditions and they have not read what we KNOW about the oral traditions from actual studies… There are tendencies in the oral traditions to elevate the status of Jesus, how they were being told in Matthew’s community would be different than how they were told in John’s community.”

2 TIMOTHY 3:16-17, (NKJV), “All scripture is given by inspiration of God, and is profitable for doctrine, for reproof, for correction, for instruction in righteousness: That the man of God may be perfect, thoroughly furnished unto all good works.”

How can any reasonably intelligent person take this seriously? A book from the New Testament of which we know was written by someone falsely claiming to be Apostle Paul, says that “all scripture is given by inspiration of God.”

If you believe something in “faith” you are more likely believing a lie.

Apostle Paul’s agenda all along was to create a religious sect off of his upbringing in Judaism but one that can ignore all of the requirements of Judaism.

> “Licentious” definition, (*The Parkway Church*) “What is licentiousness in the Bible? Licentiousness is where you allow what God forbids and think that it is okay to

> walk in sin. The word "license" in the term "licentiousness" refers to the "freedom" one thinks they have to walk in evil."

This is what evangelical Christianity has been teaching and preaching since Apostle Paul decided to make Jesus' crucifixion a sacrifice to a god. I know that there is evidence of early church fathers writing about some of the Jewish Christian sects who also believed in the sacrifice-for-sin doctrine. However, it does not feel like something that any Jewish Christian in the first century/second century era would believe, knowing how fundamentally Jewish they were. The Jewish War with the Roman Empire didn't really end until about a hundred years later with Judaism becoming illegal to practice in Jerusalem. Fundamental Judaism in the first century, did not believe in the immortality of the soul/spirit or going to any kind of heaven or hell when you die. It changed to that later from the belief in a physical resurrection from the dead. Apparently, because eventually, the original Jewish doctrine of life and death was too Jewish even for them.

48

Parables and Instinct

Mark 4:3-8, (NIV) the Parable of the Sower. *Mark 4:26-29*, the Parable of Growing Seed. *Mark 4:30-32*, the Parable of the Mustard Seed. Mark 11:12-14, the Fig Tree Withered. *Mark 12:1-11*, The Parable of the Wicked Vinedressers. *Mark 13:28-31*, The Parable of the Fig Tree.

Merriam/Webster Dictionary's definition of "instinct", "1, A natural attitude, 2, A largely specific response to environmental stimuli without involving reason. "inheritable and unalterable tendency of an organism."

This "natural attitude" of instinct in animals and humans is basically the same. The difference is our ability to alter our responses to it with or without "reason." Has humanity been fully utilizing our ability of "reason" in dealing with others and the planet that we live on?

In fact, understanding this "parable" of Adam and Eve from the above interpretation requires using the "reason" that was thrown away to replace with "faith." It is a self-deception to require "faith" to find a possibly real truth. The path to enlightenment is agnostic. Once I acknowledged to myself that

I did not know whether there is a God or truth or anything related to that subject, it was liberating and allowed me to look for evidence from anywhere and everywhere that it might be written including the pieces of the puzzle that began falling in place by logic and common sense.

Yes, logically, it required that human ability to "reason" and to question all assumptions that represented a conscious self-awareness. It is the conscious self-awareness of intelligence that the story is talking about. That sapient, sentient, conscious self-awareness of being "naked" which the parable of Adam and Eve is referring to.

"*HOW JESUS BECAME GOD: The exaltation of a Jewish Preacher from Galilee*", by Bart D. Ehrman,

> Pg. 109, [what early Christians believed] "people gain eternal life not because they believe in Christ (they have never seen or heard of the Son of Man), but because they have done good things for people in need."

Like Jesus repeatedly stressed! No matter what you believe. The Prosperity Doctrine of present day Christianity is as much of a lie as the immortal soul doctrine.

In Genesis it doesn't say that God created man. That's not what it says. It says that God said, "Let us make man in our image." Man was already here. Where the Hebrew has the words "man" and "Adam" they are not the same thing. And who's us? "Let us make man in our image, after our likeness."

According to the story, the forbidden fruit gave them the self-awareness of being naked, no longer like animals in that sense. It also gave them a cognitive mind that was necessary to obtain in our evolution from primates. However, taking the forbidden fruit was a serious violation, and a logical one to take seriously, if the purpose was to test them with a choice between feeding off of the tree of life and "live forever," or make up their own knowledge of good and evil and way of life but then die when their bodies wore out.

49

Did Jesus Exist?

Wikipedia: “Apart from a handful of references in the synoptic Gospels, the main sources for the life of James the Just are the Pauline epistles, the Acts of the Apostles, Josephus, Eusebius and Jerome, the last two of which also quote the early Christian chronicler Hegesippus and Epiphanius.”

Wikipedia: “Eusebius records that Clement of Alexandria related, "This James, whom the people of old called the Just because of his outstanding virtue, was the first, as the record tells us, to be elected to the episcopal throne of the Jerusalem church." Other epithets are "James the brother of the Lord, surnamed the Just," and "James the Righteous".”

James was appointed to lead the “Jerusalem church” immediately after Jesus’ crucifixion strictly because he was Jesus’ brother and the next in the lineage of King David. The New Testament tried to hide that completely, why? Because of Saul/Paul’s influence over his Greek and Roman followers wanting to eliminate everything Jewish about their Christology, something that James was well known for being righteous and Just. That is why the story of Paul and James coming to an agreement is so obviously fabricated. Not by what was written but by what happened after those days had passed when Jewish Christians were targets of persecution, arrested by Roman soldiers, harassed by Orthodox heretic-hunters, and singled out to be fodder for the gladiatorial games. The original Christology of a real Jesus was cut out of history by those who favored Paul's Christology for it's ease in which one can be saved and because of the Jewish war scattering those who would remain true to the original Jesus movement.

Who "sent" Paul off to be the Roman and Greek representative of Jesus? Who told him to bring the Gospel to the Gentiles? Himself! Paul claimed seeing a light in the sky when he was traveling to where he believed to find the disciples, temporarily blinding him, and a voice asking him "Why do you persecute Me?" I don't believe him or the passage in Acts which describes it. It doesn't make any sense.

According to all of the Gospels, Jesus had eleven disciples to contact who knew him and knew what he taught, but chose a Roman citizen and Greek scholar of Jewish descent who never met Jesus and was involved in the stoning to death of one of the disciples. Saul/Paul was sent by Roman authorities and the Temple priest to disperse the Jesus sect, but then created his own Christological sect in which he wanted to make it okay to break the laws and statutes of his Jewish God over something as lame as the concept of "Salvation" from the wages of sin though the primitive concept of sacrificing a human to a god.

That is what happened. It is the truth, whether anyone wants to accept it as such.

You also have to consider that the person who replaced Jesus as head of the "church in Jerusalem" was in reality the very brother of Jesus, the brother called "James," who historians and later "church fathers" of that era reveals was Jesus' real brother but the entire New Testament does not mention that with the exception of a short mention of James in one of the Epistles of Paul. Also consider the claim by Paul himself that he was actually persecuting the disciples and followers of Jesus for what might be years before his self-acclaimed "conversion." If anyone else had approached the disciples and told them that story, what would they think? However, this was the man who had been hunting down Jesus' disciples and "apostles" according to one of his own letters and the book of Acts. They would have feared Paul/Saul to the extent of being cautiously optimistic and wary of his intentions. Were Paul and Acts were both lying about Paul's persecution of the disciples? Jews couldn't arrest Jews in the fashion that Paul claims during the Roman occupation. Contrary to the claims that Paul was doing that, he wasn't arresting anyone and putting them in prison. Why lie about that? Nowhere does it explain why Paul was "hunting" down and "jailing" the "disciples" or where he would have been taking them for imprisonment. The reader is loosely led to assume that Paul/Saul was persecuting them for blasphemy against God. However, Paul was a Roman citizen who loved the Greek people more than his own, who was protecting the Roman occupation of Judea, so he would have been against any apocalyptic messianic sect that wanted Rome out of their country. He would have been persecuting the disciples because of their apocalyptic messianic sect wanting to get rid of the Roman Empire from their land.

That is why he told the disciples that he would be the emissary, the Apostle, sent out to evangelize to the Gentiles with his version of a Christology — if he said anything to them at all about it beforehand. Which I doubt he did. It was not part of Jewish Christian doctrine to convert anyone at all. Paul's argument hinged on them believing that the kingdom of God coming "soon" and Paul's idea would be to prepare for that by telling the Gentiles. To the disciples and James, this was not their concern. No indication exists historically that while Jesus was still alive that he counseled the disciples on what to do about the Gentiles in preparation for the kingdom of God. Nothing! The fact that Acts makes an effort to show Peter's strange vision of Jesus telling him to accept Gentiles into their cult means that a real Jesus never spoke to them about that at all while living.

Any passage in the Gospels where Jesus is telling his disciples to go to the people and spread the Gospel is a lie. In particular, the last verses in Mark. That was the number one agenda of what the authors of the New Test tried to push and to hide the truth about it as much as they could. It was about spreading a lucrative religious sect, profitable to those who would lead it. We should know by now, everything that seems to be hidden from the public is all about making money, power, and influence.

Even though the book called "Did Jesus Exist: The Historical Argument for Jesus of Nazareth" by Bart D. Ehrman convinced me even more that the very human person called Jesus, a transliteration of "Yeshu" or "Yehoshua," and widely known after his death much later as "Jesus", actually did exist. But I still had some doubts for a long time.

When it comes to accusations by pagans against Jesus in the years and decades following the Jewish historian Josephus it was not that Jesus didn't exist but that he may have been…

> (*Did Jesus Exist,* pg. 62) "born out of wedlock to a peasant Jewish woman who was seduced by a Roman soldier; that he was an unskilled carpenter; that he could not control his temper; and that he died a shameful death on the cross."

Aramaic speaking Jews in Jesus's native land were telling stories about him well before Paul wrote his letters after the 50 years of the Common Era (CE). If the goal of a historian is to find a wide ranging collection of people in a wide area writing about basically the same event, describing the same things about a man who was killed, attracting attention to a Jewish/Roman citizen like Paul/Saul who then persecuted those who were spreading the news of that same event and then made up his own religion from it while people were still around like James, Peter, and other of Jesus' followers — then we must concede that the historical Jesus of Nazareth (or Jesus the Nazarene) was a real man from a Jewish apocalyptic messianic sect who got caught up in a very unique situation that got him killed and later became a theology that turned into a powerful evangelical mythology that became conquerors all over the planet and became one of the three major religions.

Eventually, all of the Jewish Holy Days were covered up or just excluded from Catholic Christian tradition. The Saturday Sabbath, the seventh day, mutated into a Sunday/Sun-god, worshiping Catholic Christian day of rest and worship, completely against the fourth commandment. Passover became Easter, which celebrated a pagan Greek goddess of fertility with eggs and rabbits. The feast of unleavened bread which lasts for a week after the Passover was completely abandoned by the "licentious" Christology that Paul started which became the only Christology left and mutated into the established Christian "Catholic" religion of the Roman Empire. The Jewish Feast of Weeks after Passover was taken over by Christian "Pentecost, (Greek for, fifty days)" as a celebration of Jesus leaving his disciples for the last time and miraculously rising up into the sky; surely a purely made-up event that never happened and not a historically recorded event at all. So, out of the 7 (including the Jewish "Day of Trumpets," the "Feast of Tabernacles," and the "Last Great Day") annual Holy Days from the Hebrew Bible (not the later Jewish holidays) the "Church" that dominated over all other Christologies pushed out everything Jewish about Jesus' apocalyptic messianic Jewish sect and became Catholic ("universal and triumphant") about just the life and death of Jesus with man-made mythological doctrines thrown in. It was no longer Jewish in any way and those who tried to follow any kind of Jewish Christianity in the wake of the Jewish war against Rome was so much more persecuted than those who followed Paul's or other's Christologies. It was quite a wave of destruction to the Jewish people and their future, to their faith, and to the original Jewish Christology!

Ehrman says that the earliest Jewish followers were around Jesus at the same time that Jesus was crucified but fled afterwards! Doesn't logic tell us that they would know more than anyone who Jesus was or thought that he was and why? Of course, but religion has been hiding those facts. Why? Because Jesus was one of the leaders of a purely Jewish sect with Jewish beliefs and a socialist in his ideology. His followers were not what Christianity later became to spread a "gospel" industry of mass appeal, pandering to a mass delusion for profit.

Ehrman knows also that the "New Testament is filled with discrepancies and contradictions in matters both large and small."

(Did Jesus Exist?) "The Gospels contain non-historical materials."

Mark 15:6-15, "When Jesus was on trial, Pilate offered to release one of his two chief prisoners, Barabas or Jesus, as was allegedly his custom at Passover."

As stated above, we have no historical record of any such custom being conducted by Pilate or any other Roman governor. In fact, we have historical evidence that this was not the case at all, especially from the real historical "Pontius Pilate."

> (*Did Jesus Exist?*)"And it defies imagination that the historically ruthless Pilate, not known for currying favor among the crowds, would be willing to release a violent and dangerous insurrectionist every year just because crowds wanted him to do so."

It didn't happen! We know that for sure. It was fictitious, and we know why they made it up.

Again, the authors of the Gospels were not historians but what they were was Greeks who lived at a time when Roman soldiers ransacked and destroyed Jerusalem because of the revolt by the Jewish people against the Romans from 66 to 135. In this case, the Gospel writers were obviously trying to put as much blame, as much fault, on the Jewish race for the crucifixion of Jesus and therefore away from the Romans who would not look kindly toward any religious sect that vilified them even though the Romans were, in fact, the ones who ultimately tried, convicted, and carried out the crucifixion.

Pilate was later charged with cruelty for having men executed without a trial. If we know now that the story of Pilate trying to release Jesus against the apparent wishes of the Jewish people was a lie that all four Gospel writers wrote about and that Pilate was charged with cruelty toward civilians by the Emperor Caligula who ordered Pilate to commit suicide, how much of what happened according to the Gospels has any real truth to it? Why was Jesus crucified? Some would say that it was because of his outburst at the Temple with the "money-changers." However, the size of the Temple was so huge that he would have been just kicked out and any disturbance like that would not necessarily lead to a death penalty like a crucifixion. There is much that we cannot know about what happened to Jesus just based on the New Testament. There was other Gospels of all kinds that were considered for inclusion into the NT but only those four were chosen. What would be the agenda of those who just wanted those four Gospels to be preserved? All of the reasons I have discussed would be logical.

And we should know by now that the tragedy of blaming Jewish people for anything and everything throughout history can definitively trace back to the Gospels and the religion that became the "Holy Roman Empire."

If you'll notice something going forward, it is that Conservative Christian theology has always been vulnerable to becoming the political tool used for embracing authoritarian, narcissistic, genocidal, tyrants like the OT "God" while ignoring what the Jesus of the Gospels is quoted (whether historically of not) saying and what he was doing. History has many instances where Jewish people and entire villages were blamed for things from a failed economy to plagues, crop failures, destructive weather events and other natural disasters. Why? Think about it. Christianity ruled most of the "known" world after "Explorers" and "Colonialism" ravished countless Indigenous people around the world to force the Christian religion on them and steal their gold. They forced their form of the Christian religion upon the entire world of which they could sail to, and very much did so, carrying their cross — with their armor and sword — to Christianize the world in their own image by force. With them came antisemitism, as they brought their Bibles with "missionaries" to the four corners of the world. Missionaries who believed that primitive people had no rights, and all women and children were the property of men. Do you see the pattern? Do you see what happened? To quote a standard phrase attached to simple observations, it isn't rocket science.

Finally, of whether Jesus existed or not we are faced with the most perplexing issue of all, if he did exist, who and what did Jesus think he was? There is sufficient evidence and logic that he did exist. It is an understanding of the original Jewish Christians where we can see Jesus as a messiah wannabee and could have believed he was the second Adam.

The real Jesus was probably a Jewish Teacher of messianic apocalyptic persuasion who believed that being baptized gave him the fruit from the tree of life and "Holy Spirit," and that he would lead the way to Armageddon at the same time as growing his God-seed. Unfortunately for him, it did not work at all as far as we know now from historical facts and common-sense logic. There wasn't any Armageddon. No messianic apocalyptic event ever took place. No divine intervention. There most likely won't be one in the conventional religious sense because we are in a process of evolving which requires non-interference by the Designer. Was it Jesus's intention to bring about the defeat of the Romans and free Judea for self-governance under the same "God" as his ancestors? We may never really know the answer to that. However, the concept of his endeavor to be the "second Adam" is a real possibility when we look at what he did following his baptism by John the Baptist. Not necessarily based on what the Gospels say that he said and did, they tried not to include the beliefs of the Jewish Christians. But it is logical just by what his actions were that lead from his baptism to his crucifixion and the inclusion of John the Baptist by the Gospel writers. It fills in the blanks that scholars have noticed for generations but were afraid to reveal because of the authoritarian regimes of Christianity.

Reading the four different accounts of the crucifixion of Jesus, one can't help wondering why the details are so different from each other. It looks like each Gospel writer was writing about the events that occurred but making stark differences in the details. Essentially, just making it up as they went. That is what it appears to be. Most of what is written in all the Gospels are fabricated or embellished accounts. If you need to hear this from the words of Professor Ehrman himself here it is...

> *DID JESUS EXIST?* Page 268, "… we do not know who the authors of the Gospels actually were other than that they were almost certainly not the Aramaic-speaking lower class who made up the early disciples of Jesus." "… the different Gospels present very different portrayals of who Jesus was, what he stood for, and what he preached…" "... the New Testament tales of Jesus are full of discrepancies in matters both large and small."

To claim that tall tales about Jesus is how mythologies of the ancient world invented the "myth" of Jesus denies the fact that the first century Jews were strict monotheists. The Jewish people of that era would not have made a human being appear to be like their god in any way. They believed that there is only one god of importance, their god, and that god can only be one god — not two, not three, and not multiple layers of a heavenly realm like in Greek mythology. Later Christianity mutated towards polytheism with its Trinity doctrine and the addition of "Saints" in heaven. It is all very polytheistic, pagan, and Gentile views; meaning, more than one god and several eternal entities of humans becoming the so-called "Saints" would be like becoming a god themselves to ancient people. Centuries later they had Catholics praying to "Saints" in heaven or Mother Mary instead of God. This form of polytheism was comparable to and compatible with the Greek and Roman mythological religious beliefs back then, and the pagan immortal soul doctrine, as all other religions of that time were polytheistic. Not the Jewish religion though. They were adamant toward insisting in there being only one god.

Ehrman, "pagans recognized many gods living in all sorts of places and serving all sorts of functions. The God of the Jews was believed by the Jews (and them alone) to have created the world and ultimately to be sovereign over it. Jews did not insist that other people worship this God, but he was the only God for them."

What if the body of Jesus was not really taken down and put in a tomb? Is there evidence of that by critical scholars?

> "*How Jesus Became God: The Exaltation of a Jewish Preacher from Galilee*" by Bart D. Ehrman, (pg.160) "In sum, the common Roman practice was to allow the

> bodies of crucified people to decompose on the cross and be attacked by scavengers as part of the disincentive for crime. I have not run across any contrary indications in any ancient source. It is always possible that an exception was made, of course. But it must be remembered that the Christian storytellers who indicated that Jesus was an exception to the rule had an extremely compelling reason to do so. If Jesus had not been buried, his tomb could not be declared empty."

The Gospel writers composed their stories decades after the events that they narrate. No one was taking notes when Jesus was alive. Ehrman's book on this subject is absolutely amazing in revealing insightful common-sense answers about the historical events that began Christianity. The crucifixion of Jesus and alleged resurrection is the defining moments. If there are too many holes, contradictions, historical inaccuracies, falsehoods, and outright absurdity in the writings that Christianity deems sacred and true we must accept the truth and reevaluate what we put our faith in.

50

Cognitive Dissonance

The God of the 3 major religions, evolved long ago by humanity to be beyond the verification of science. Before the discovery of the sciences revealed religion's fictitious claims, science did not exist at all. Your religion was your science as well as your politics. There was no distinction between the two, they were considered the same. You did not need to "believe" in "faith" what everyone thought they knew was absolutely true. Very few individuals were skeptical and even fewer were brave enough to question their indoctrinated religion openly. So, the vast majority of people were believers because there was no evidence yet to show that religion was wrong about anything. No reason to doubt other than that nagging conscience of uneasiness at religion's authoritarian/totalitarian misogynistic, racist, bigoted, and genocidal tendencies.

Before critical scholars informed religions that they did not have their facts right, people did not need to talk about needing "faith." Their primitive traditions of beliefs handed down from their ancestors were the limits to their collective knowledge for ages until recent centuries. They didn't need faith in a religion when religion was all there was. By claiming that it requires only "faith" puts the existence of God beyond the verification of science for the devout believers and use it for filling collection plates. The idea of faith protects the idea of God from disproof. It is a convincing deception to many. It has caused a mental condition of discomfort and denial in countless people of faith called "Cognitive Dissonance."

We know that Jesus taught in the Gospels, and probably in real life as well, that giving your life over to God actually means being the servant of everyone, not a worshiper or servant of God. However, there is a great deal of cognitive dissonance among today's Christians who follow the Prosperity Doctrine. When delusions become reality in the minds of the faithful, there's a level of mental anguish of cognitive dissonance to those who know what Jesus says against wealth and comfort for oneself, but they use their God-given prosperity to be selfish, opinionated, and self-righteous.

They must then ignore those parts of the Bible that teaches servitude to others as their conservative political ideology ignores the homeless, the destitute, and the indigent — and still believe that they function as believers with prayers, celebrating Christian traditions and holidays and even being comforted by a religious theology that practically guarantees an eternal life in "paradise" without doing practically anything for it — holding on desperately to their elitist apathy toward those who suffer from abject poverty. Exactly the opposite of what Jesus taught in every Way!

> *ACTS 9:1-2* (NKJV), "Then Saul, still breathing threats and murder against the disciples of the Lord, went to the high priest and asked letters from him to the synagogues of Damascus, so that if he found any who were of the Way, whether men or women, he might bring them bound to Jerusalem."

What is that all about? First of all, why is Saul "breathing threats and murder" against Jesus' disciples? No one has a good answer for that. By "high priest" from Jerusalem, it can only mean the high priest of the Temple. However, the main issue with this is how the high priest or Saul had no authority to have anyone arrested or bound, and no place to put them into incarceration, or hold a trial of any kind, under Roman occupation.

The Romans had full control over the populations of their empire. It just wasn't how things were done. So who is this author writing to? It would be for Greek and Roman citizens of the growing Pauline Christology near the end of the Jewish war around 120 CE. The average citizen of the empire and follower of mythological religions like they were would not know or care. The intention is to show Saul/Paul having authority to hunt down and arrest people under strange circumstances. However, it is possible, as I suspect it was true, that Saul/Paul was against the apocalyptic messianic sects of Judaism who were fighting for freedom in a war that started only a couple decades after the crucifixion. Anyone who was hoping for a messiah to come and defeat the Romans was seen by Paul as making trouble for everyone's stability, as reflected in the conflict that was boiling up at the time. Pilate would not hesitate to have anyone executed in the most horrific way devised if that person was a possible leader of a messianic group — which Jesus at the time could be associated as regardless of who he was or what he said, even a righteous and holy Teacher. Also, what does this author mean by calling the disciples "the Way"? If I'm not mistaken, I believe one of Saul/Paul's own Epistles have him admitting that he "persecuted the Way." We can't use the book called Acts or the Gospels as a reliable source of what happened.

> (*WIKIPEDIA*) "Cognitive dissonance is the perception of contradictory information and the mental toll of it. Relevant items of information include a person's actions, feelings, ideas, beliefs, values… Cognitive dissonance is typically experienced as psychological stress when participating in an action that goes against one or more of those things. According to this, when two actions or ideas are not psychologically consistent with each other, people do all in their power to change them until they become consistent. The discomfort is triggered by the person's belief clashing with latest information perceived, wherein the individual tries to find a way to resolve the contradiction to reduce their discomfort."

When Christians learned that the planet we live on is billions of years old, and that all life evolved from the same common ancestors, they cannot justify calling it "true" because of their indoctrination in the Biblical Creation week. This is why they insist on teaching their children that dinosaurs were alive at the same time as people. Their cognitive dissonance looks for any excuse to keep believing what their religion taught, even when it is revealed to them that it is wrong. Their excuse for extinct species like dinosaurs is that they were not included on Noah's Ark and therefore, went extinct because of that. This way, they remove the discomfort of learning the scientific facts that annuls their beliefs.

Whose God is the right one? None of them! The only real GOD, if there is one, would be benevolent, passive, empathetic, silent, absent, and uninvolved in human affairs, a life-form that is somehow outside of our dimensions who has not been involved with human beings at all.

The following verses are from the New International Version. What does an evolved conscience think about this...

Exodus 21:12 "Anyone who strikes a person with a fatal blow is to be put to death."

Leviticus 24:17-22 "And whoever kills another person must be put to death.

Exodus 22:19 "Whoever lies with an animal shall be put to death.

Leviticus 20:10,27 "'If a man commits adultery with another man's wife — with the wife of his neighbor — both the adulterer and the adulteress are to be put to death.' (27) 'A man or woman who is a medium or spiritist among you must be put to death. You are to stone them; their blood will be on their own heads.'"

Exodus 22:2-3 "If a thief is caught breaking in at night and is struck a fatal blow, the defender is not guilty of bloodshed; but if it happens after sunrise, the defender is guilty of bloodshed.

1 Samuel 15:3 "Now go, attack the Amalekites, and totally destroy all that belongs to them. Do not spare them; put to death men and women, children and infants, cattle and sheep, camels, and donkeys.'"

Romans 13:1-5 (written by Paul) "Let everyone be subject to the governing authorities, for there is no authority except that which God has established. The authorities that exist have been established by God. Consequently, whoever rebels against the authority is rebelling against what God has instituted, and those who do so will bring judgment on themselves. For rulers hold no terror for those who do right, but for those who do wrong. Do you want to be free from fear of the one in authority? Then do what is right and you will be commended. For the one in authority is God's servant for your good. But if you do wrong, be afraid, for rulers do not bear the sword for no reason. They are God's servants, agents of wrath to bring punishment on the wrongdoer. Therefore, it is necessary to submit to the authorities, not only because of possible punishment but also as a matter of conscience."

It's easy to see from that passage why Apostle Paul passionately defended the Roman occupation and pushed for his Christology to be as far from the original messianic, apocalyptic Jewish sect as he could take it. Believing that a messiah was coming to defeat the Romans was a doctrine that he fought against, but it was exactly what Jesus's sect was all about. That is the real reason he persecuted the disciples by his own admission and had a hand in the stoning of Steven. I suppose Paul/Saul would say that Hitler was one of God's servants too if all "authorities that exist have been established by God." Really! He was serious. It's absurd, and yet, we know that it came from the mind of Apostle Paul himself and is not disputed by Christianity because of their faith in all their holy writ. Once again, cognitive dissonance tries to justify those statements any way possible.

Deuteronomy. 33:27, "The eternal God is thy refuge, and underneath are the everlasting arms: and he shall thrust out the enemy from before thee; and shall say, Destroy them."

Really? That "God" of theirs is NOT a benevolent being. That was written by demented human minds. It couldn't be anything else.

God supposedly said to Abraham...

Genesis 22:17, (NRSV) "I will indeed bless you, and I will make your offspring as numerous as the stars of heaven and as the sand that is on the seashore. And your offspring shall possess the gate of their enemies."

He also told Abraham to kill his son and then stopped him from doing so.

Bible Verses about Destroying Enemies: lots of them.

Romans 16:20, (NRSV) "The God of peace will shortly crush Satan under your feet. The grace of our Lord Jesus Christ be with you."

Satan is a completely and thoroughly fictitious character.

2 Chronicles 15:12-13 (ESV) And they entered into a covenant to seek the Lord, the God of their fathers, with all their heart and with all their soul, but that whoever would not seek the Lord, the God of Israel, should be put to death, whether young or old, man or woman."

Luke 19:27 (ESV), "But as for these enemies of mine, who did not want me to reign over them, bring them here and slaughter them before me.'"

Matthew 10:34 (NRSV) "Do not think that I have come to bring peace to the earth. I have not come to bring peace, but a sword."

In all of the Gospels and much of the New Testament as a whole we find passages that are added by the authors in order to make Jesus' God appear to be the same old violent God from ancient Judaism. There are also obvious embellishments to make Jesus appear to be born divine, or divine from the beginning of time, or rebellious against the Jewish laws and authority. The Gospel writers were trying to paint a picture of a vengeful authoritarian God of Jesus in contrast to the one that we are fairly sure Jesus talked about. A God of peace and altruism.

There is that cognitive dissonance again. Christians read where God is a vengeful and violent personage and then read that Jesus stood for peace and being of service to those in need. How do they account for such a wide difference? By making stuff up or ignoring the parts that conflict with the belief of a petty, vengeful, and intolerant God.

> *Bart D Ehrman*, "The Gospels of the New Testament don't claim to be written by eyewitnesses and in fact they were not written by eyewitnesses. The Gospel writers were living forty, to fifty, to sixty years after Jesus died. They wrote the Gospels in Greek; Jesus' language was Aramaic. These Gospel writers were living in a different country decades later. Where did they get their information from? They were not the followers of Jesus, they don't claim to be followers of Jesus, the disciples, they were written by people decades later in a different language. Where did they get their information from? They heard stories about Jesus that had been in circulation year after year after year, decade after decade down to the time that the Gospel writers living in a different country, speaking in a different language heard the stories. What happens to all stories when they are transmitted orally? They change. The Gospel writers have discrepancies among themselves because the stories that were told and retold were changed over time and the Gospel writers themselves sometimes changed the stories. That's why there are discrepancies, that's why scholars might be able to tell you generally what Jesus was all about. They can list 8 things that Jesus did, but they can't tell you the details and agree. Why can't scholars agree? Because there are so many discrepancies that the Gospels are not reliable."

51

The Sin-Day Sabbath

> *EXODUS 20:8* (NKJV), "Remember the Sabbath day, to keep it holy. Six days you shall labor and do all your work, but the seventh day is the Sabbath of the LORD your God. In it you shall do no work: you nor your son, nor your daughter, nor your male servant, nor your female servant, nor your cattle, nor your stranger who is within your gates. For in six days the LORD made the heavens and the earth, the sea, and all that is in them, and rested the seventh day and hallowed it."

So which day of the week should be the "Christian" Sabbath? It never changed from Saturday to Sunday until the Emperor Constantine wanted to blend the new Christian religion with the old Roman and Greek traditions. That's not what the historical Jesus, or for that matter the Jesus of the Gospels, would have done at all. "Keep the Commandments" was something he is quoted saying often. Some of the Greco-Roman religions were designated to worship a god of the sun. They worshiped their deities of the sun on the first day of the week on the same calendar everyone uses today. The Jews had always (still do) kept the "seventh day" of the week, which on your calendar is still Saturday and always has been. The earliest "Christians" kept the Jewish Sabbath day as well, even Apostle Paul (according to his letters and ACTS). The New Testament has no quotation of any kind that distinctly changes the "Sabbath" to Sunday or even implies that God or Jesus changed it. The form of Christianity that chose to make Sunday their day of rest and worship to their "God" has tried to use any place in the New Testament that they could find to twist and spin the interpretation into an excuse for calling Sunday the "Lord's Day."

It's a total scam and deception, like so much of Christianity, and it makes me nauseous to know that so many people who have called themselves Christians never really questioned why their "day" of rest is Sunday instead of the day that Jesus and all Jews have kept. I hate it when people say that it doesn't matter. They obviously don't know what they're talking about.

> *EXODUS 20:8-11*, (GNB, the "Fourth Commandment") "Observe the Sabbath and keep it holy. You have six days in which to do your work, but the seventh day is a day of rest dedicated to me. On that day no one is to work — neither you, your children, your slaves, your animals, nor the foreigners who live in your country. In six days I, the LORD, made the earth, the sky, the seas, and everything in them, but on the seventh day I rested. That is why I, the LORD, blessed the Sabbath and made it holy."

Back in the old days, before even modern timekeeping was invented like clocks, each day had "a morning and an evening, one day", or as it states in the Good News Bible...

GENESIS 1:4-5, (GNB) "God was pleased with what he saw. Then he separated the light from the darkness, and he named the light "Day" and the darkness "Night." Evening passed and morning came — that was the first day."

NOTICE that he separated the night from the day and the night was a kind of Passover to the next day. They were not equal. They were separate. Night was a passing on to the next day.

Our 24-hour period consists of 12 hours of light followed by 12 hours of darkness. But the darkness was a kind of "Passover" to the next day. It "passed." The beginning of each "Day" was at first light of each day, "and morning came", not at midnight, nor at sundown. A "day" was technically the hours of daylight. Their way of saying what time it was, was in the amount of hours in the day which they distinguished by noticing where the sun was and using sundials. It wasn't until the Jewish people were enslaved in Babylon that they adopted the 24-hour period and included the "darkness" as part of a full day. From that point on it was the beginning of the 24-hour day at sundown rather than at sunrise.

The main reason popular Christianity has been keeping the Sunday Sabbath, which could be called the "Sin-day Sabbath", is because the Roman Emperor Constantine who declared Christianity to become the official religion of Rome insisted that their day of rest and worship must not be the Jewish

Sabbath but the day they held sacred under the Sun-god deities. Today make an excuse for the change to Sunday, the Church claimed it is the day that "Jesus" supposedly was raised up. However, even if one uses the New Testament accounts of that morning's event at the tomb where Jesus's body was supposedly laid you will see that in each case Jesus was not there and had already "risen" as the "dawn" of the "first day of the week" was approaching.

MATTHEW 28:1-2,5-6, (GNB) "After the Sabbath, as Sunday morning was dawning, Mary Magdalene and the other Mary went to look at the tomb. Suddenly there was a violent earthquake; an angel of the Lord came down from heaven, rolled the stone away, and sat on it... The angel spoke to the women. "You must not be afraid," he said. "I know you are looking for Jesus, who was crucified. He is not here; he has been raised, just as he said. Come here and see the place where he was lying."

Jesus had supposedly maintained that he would be in "the grave" for three days and nights.

So, this "Gospel" of the New Testament claims that on that early Sunday morning, as morning "was dawning", an angel opened up the tomb and told them that Jesus was already "risen" from the dead, "has been raised", and was no longer there. Other translations quote it deceptively as, "he is risen."

Let's see what the earliest Gospel written says…

MARK 16:1-8, (GNB) "After the Sabbath was over, Mary Magdalene, Mary the mother of James, and Salome brought spices to go and anoint the body of Jesus. Very early on Sunday morning, at sunrise, they went to the tomb. On the way they said to one another, "Who will roll away the stone for us from the entrance to the tomb?" [It was a very large stone.] Then they looked up and saw that the stone had already been rolled back. So they entered the tomb, where they saw a young man sitting at the right, wearing a white robe — and they were alarmed. "Don't be alarmed," he said. "I know you are looking for Jesus of Nazareth, who was crucified. He is not here — he has been raised! Look, here is the place where he was placed. Now go and give this message to his disciples, including Peter: 'He is going to Galilee ahead of you; there you will see him, just as he told you.'" So they went out and ran from the tomb, distressed, and terrified. They said nothing to anyone because they were afraid."

That was the end of the oldest documents of that text, which was what the original said according to all accredited scholars. Jesus was not there. Notice also that it started by stating

"After the Sabbath was over." It was Sunday morning, or as LUKE calls it, (New King James Version, NKJV). *LUKE 24:1*, "Now on the first day of the week, very early in the morning…" And John's Gospel has it, *JOHN 20:1*, "Now on the first day of the week Mary Magdalene went to the tomb early, while it was still dark, and saw that the stone had been taken away from the tomb."

All four Gospels agree that the day of the week was the "first day of the week," early, and Jesus was already gone. Everything else they stated after that is quite different in each

Gospel. They had a name for the seventh day of the week, they called it "Sabbath" or "The Sabbath," but every other day of the week had no name. They just called it "the first day", "the second day", "the third day", and so on. According to this, Jesus did not come back to life on Sunday. It happened before that morning.

> *Britannica:* "It appears that the notion of the Sabbath as a holy day of rest, linking God to his people and recurring every seventh day, was unique to ancient Israel."

The other religions were mostly sun-god worshipers and rested on the first day of the week if at all. Mark was the oldest of the Gospels written, estimated to have been written only forty years after the crucifixion. A letter (epistle, 1 THESSALONIANS) by Paul included in the New Testament is the oldest one of the New Testament books. The other writers of the Gospels were written more than three decades later, and all have quite different accounts of what happened that Sunday morning with many varied details. That alone should raise a flag of skepticism about the resurrection itself. That first Gospel written, Mark, left the story to end with the group who went to the tomb leaving the tomb, "saying nothing to anyone, because they were afraid." The last 11 verses of that last chapter of the text in the vast majority of translations was a later addition to the text by Orthodox scribes and those copies ended up being copied over and over until all of Christendom had no idea that they were reading someone's words at the end of Mark that wasn't in the original. Furthermore, no one knew for well over a millennium that this was the case and that the character "Mark" labeled as the author was not the Mark from the Gospels, it was labeled with that name decades later to distinguish it from the other Gospels going around, and no one knows who actually wrote any of the Gospels. In addition to that, every book in the New Test has gone through changes as it was passed down from generation to generation, century to century. What we end up with now is a book that is almost certainly human inspired, man-made, written by people not associated with it —pretending to be written by someone from the Gospel stories — and very "fallible."

If a real God had wanted "His" words and influence to be in those writings, He would have preserved the originals or at least had clarifications added. But that is not what scholars have found out. Even the Hebrew Bible and the Islamic Quron was preserved by scribes much more accurately than the New Test. We know that now because we have critical thinking and cognition skills along with the centuries of research. The Holy Bible is the single most studied and researched literature ever. When I say it has been studied I am not talking about "Bible Study" by believers to see what it says about social issues.

There's plenty of that too. But in this case I mean researched to discover who might have wrote it's parts, when they were written, how they changed over time, who changed what, why, and how. That is what scholars do. I tried to join a Bible Study recently to share these and other facts, but was quickly shut down. I was told it was not the topic of their discussion for that day. Funniest thing about that was that they didn't have a clue of what I was talking about and wasn't interested.

Christians don't want to acknowledge that their sacred writings are the most researched and investigated texts in history. The only reason that we know about the end of Mark is because scholars have possession of legitimate documents of the text by very early Christianity showing that the story ends at *MARK 16:8*.

You should never equate anything written as coming from a god of any kind but particularly when we know that it's either a forgery or was labeled with the name of someone from the Gospel stories that was written by someone else decades after the characters from the stories were all dead.

The Gospels all account that Jesus had already come back to life before that morning. Technically, that would not include Sunday morning, which is why all four Gospel accounts put the discovery of finding Jesus' body gone on a Sunday morning. They would not go to the tomb during the weekly Sabbath or Passover to check on Jesus' body or "anoint" his body with oil because it would have been against their laws. So, the excuse for making Sunday the Christian "Sabbath" doesn't pass the test of logic.

Popular Christianity had its various ways that their "believers" can obtain a free (minus the cost of tithes and offerings) ticket to their proverbial "heaven" of mythology. They decided to change their day of worship to the day that suited the Greek and Roman religions. They changed the names of the key figures in their religious writings from Jewish names to Greek and Roman friendly ones, and finally eliminated anything "Jewish" about what was originally a very

Jewish Sect. Those are the facts. It is easy to see the Jewishness of the New Testament, particularly in the Gospels, being written out of the text if you research and study it with an open but slightly skeptical mind.

MATTHEW 12:11-12, (NRSV) "At that time Jesus went through the grain fields on the Sabbath; his disciples were hungry, and they began to pluck heads of grain and to eat. When the Pharisees saw it, they said to him, "Look, your disciples are doing what is not lawful to do on the Sabbath." He said to them, "Have you not read what David did when he and his companions were hungry? How he entered the house of God, and they ate the bread of the Presence, which it was not lawful for him or his companions to eat, but only for the priests? Or have you not read in the law that on the Sabbath the priests in the temple break the Sabbath and yet are guiltless? I tell you, something greater than the temple is here. But if you had known what this means, 'I desire mercy and not sacrifice,' you would not have condemned the guiltless. For the Son of Man is lord of the Sabbath."

Notice that it was the "Pharisees" who confronted Jesus on this. There is evidence that Jesus was a part of the Pharisees at some point. It was the Greek writers of the Gospels making it look like they were against Jesus or, at the least, questioning his teachings. Remember that it was the agenda of the writers of the New Testament to make it look like Jesus was eliminating the requirements of keeping Jewish laws and statutes by all of God's chosen, including Paul's Greek and Roman friends. Apostle Paul is even quoted saying that he is above the law, which you will see for yourself coming up.

Think for a second about when the God of the Old Testament ordered the Hebrews to slaughter the Canaanites and take their land. The God of the Old Testament was traditionally believed in the Hebrew Bible to be a strict authoritarian who made Commandments that could get you killed for if you did not obey them to God's satisfaction. Sometimes they were even told to kill everything in a village except the young ladies, those they could take for themselves. In one verse from the Old Testament has someone saying…

PSALM 137:9, (NRSV) "Happy shall they be who take your little ones and dash them against the rock!"

Where are the anti-abortion Christians? Anyone of the "faithful," the "believers," who would willingly worship a religion like that must be suffering from extreme cognitive dissonance unless they are sociopaths. And likely they never knew any of this. Most history books, as well as Christian institutions, downplay the role Christianity had in the history of war in order to hide atrocities including the Holocaust, "The Spanish Inquisition," the "Crusades," the "Knights Templar," and the Spanish Conquerors who committed genocide around the world in their ambitions to colonize and convert everywhere with their religious belief that God was working through them.

However, and here's the catch, there probably was no tomb that Jesus was put in to begin with. The tradition was to leave the bodies on the cross for days. In the final analysis of the resurrection of Jesus, there is sufficient reason to believe that Jesus did not come back to life.

So, the excuse to change the Sabbath to Sunday because Jesus came back from the dead on a literal "early" Sunday morning is a totally false premise according to the Holy Bible itself and historical analysis.

One of the false claims of Sunday Christians is that Jesus disagreed with the "Pharisees" that the Sabbath should be kept at all. However, any close examination of those alleged incidents from the Gospels clearly reveals that Jesus did not argue about whether the Sabbath should be kept, but how it should be kept, and those accounts are probably fabrications by the Greek writers or forged inserts by scribes into already existing documents before making copies by hand, which was later copied with the changes made. The Christology that became the "Orthodox" one eventually not want anything Jewish to be considered sacred to Christians. They purposely blended Christian doctrine, tradition, and rituals to merge with the most popular pagan mythologies. All Old Testament Holy Days were considered Jewish and either eliminated or somehow changed to appear more pagan than Jewish.

Under the strict observance of the Jewish authority of Jesus' day, a doctor could not do anything to help an injured person on the seventh day of the week. Even this was a form of "work." One could not travel more than a short distance, about a mile, on the Sabbath. You could not even pick some fruit from a tree or bush or any edible plant on a Sabbath. As a "Commandment", this was the Law. Jesus knew this was barbaric, fundamentalist extremism and completely contrary to the Designer's "Way." That was the reason he "healed" people (if that was true) on the Sabbath. He was not defying the

Sabbath law, the fourth commandment, he was defying the callous inhuman enforcement of the Sabbath law by the delusional, self-righteous humans pretending to represent their God. That was not to disregard keeping the Sabbath altogether or changing it to another day of the week. He believed that he was in a process of becoming the very first offspring of God. That his God is merciful and benevolent, not the monster the ancient Judaism worshiped.

However, whether any human beings can advance to a higher level of existence would naturally require a higher sense of wisdom and benevolence. That is something that human beings extremely lacked then and still do now 2000 years later.

How does anything justify Christianity's change from the seventh day to the first? Any calendar we use today shows the first day of the week as Sunday while the other end of the calendar's chart sits the seventh day, Saturday. It was important enough to God to make it part of the ancient Creation Week, feed the Israelis in the desert twice as much food on Friday to make sure that they didn't work on Saturday, and made it one of the Ten Commandments. That appears to me that it would be important for some reason to this "God. No matter what you think about it all, this would be a very important point. If we can't get this one right, can we get anything right? Think about it.

And another thing, if we agree that the Designer does not consist of the same matter and energy of this universe and not consisting of any material form, what would be the most logical choice IT would have to give us as a symbol of

ITSELF to "remember"? What would be a symbolic "image" of ITself that we material-based and temporary lifeforms could relate to? What kind of "sign" would IT give to IT'S "chosen", for a particular stage of our evolution? One which does not change over time? Something that will not decay, or "see corruption," that cannot be mistaken for anything else? Something that does not even evolve? Something that you could mark on your calendar? Of course, a period of time itself, like a day of the week!

Over the course of human history, the biggest difference between the Jewish race and the rest of the world has been the day of the week that they solemnly keep as their day of rest and worship. Even just the concept of a day off of work was new to people back then. It was bare survival that they had to live by, it was not punching a time-clock.

EXODUS 31:16-17, "Therefore, the people of Israel shall keep the Sabbath, observing the Sabbath throughout your generations, as a perpetual covenant. It is a sign forever between me and the people of Israel..."

EXODUS 31:13, "You shall keep my Sabbaths, for this is a sign between me and you throughout your generations, that you may know that I, the LORD, sanctify you."

LUKE 4:16,31, "And he (Jesus) came to Nazareth, where he had been brought up; and he went to the synagogue, as his custom was, on the Sabbath day. And he stood up to read... And he went down to Capernaum, a city in Galilee. And he was teaching them on the Sabbath."

JOHN 6:2, "And on the Sabbath he began to teach in the synagogue..."

JOHN 7:23-24, "If on the Sabbath a man receives circumcision, so that the law of Moses may not be broken, are you angry with me because on the Sabbath I made a whole man's body well? Do not judge by appearances, but judge with right judgment." Exactly!

ACTS 15:21, "For from early generations Moses has had in every city those who preach him, for he is read every Sabbath in the synagogue."

ACTS 13:13-14,42-44, "Now Paul and his company set sail from Pathos and came to Pergia in Pamphylia. And John left them and returned to Jerusalem; but they passed on from Perga and came to Antion of Pisidia. And on the Sabbath day they went into the synagogue and sat down... As they went out, the people begged that these things might have told them the next Sabbath. And when the meeting of the synagogue broke up, many Jews and devout converts to Judaism followed Paul and Barnabas, who spoke to them and urged them to continue in the grace of God. The next Sabbath almost the whole city gathered to hear the word of God."

HEBREWS 4:9-12, "So then, there remains a sabbath rest for the people of God; for whoever enters God's rest also ceases from his labors as God did from his. Let us therefore strive to enter that rest, so that no one falls by the same sort of disobedience. For the word of God is living and active, sharper than any two-edged sword, piercing to the division of soul and spirit, of joints and marrow, and discerning the thoughts and intentions of the heart." (unknown author but not by Paul)

"Science" can be defined as a "systematized knowledge derived from observation and study."(*Wikipedia*) "Reason" can be defined as, "the ability to think, draw conclusions, etc." and "to think logically; analyze" and "The power of the mind to think, understand, and form judgments by a process of logic (*Google*)." Britannica says, "think, cogitate, reflect, speculate, deliberate means to use one's powers of conception, judgment, of inference."

We can observe and study knowledge while our ability to reason allows us to draw the most logical conclusion. Simple? It should be as simple as that. However, we must use that logic and reason rather than the "faith" that blinds our eyes away from producing complex abstract thoughts, critical thinking, and reasoning.

52

"I Myself Am Not Under The Law"

(Wikipedia) "the teachings of the Apostle Paul asserted that physical circumcision was unnecessary for the salvation of Gentiles and their membership in the New Covenant."

That is what Acts says, which was written by a Greek follower of Paul. Later in Acts it becomes obvious that Paul wanted to eliminate keeping all of the Laws and statutes of the Old Testament. Once a reading of the Epistles of Paul is studied it becomes clear that Paul was against everything about those first real followers of Jesus and therefore, against Jesus himself. Paul was the true Antichrist, if there ever was one. It was his motivation in becoming the founder of his own version of a Christology and the Greek writers of the New Testament lied to hide what he was really doing. Turning the crucifixion of Jesus into a human sacrifice to a god. You know that's what he did.

The "Apostle Paul," was the most dangerous man in history. The Jewish Christians did not follow him, they saw him as a "heretic" (one who holds an unorthodox opinion). The heading to this chapter is something Paul wrote in one of his actual Epistles but was taken out by later Scribes who made copies for preserving the text and didn't want it to say what it said in the context that he wrote it. That fact was only discovered for the first time this past century. Every "Version" of the Bible that I know of, except the NRSV, has the version that ended up in all English Bible translations to this point. They leave out that line completely to hide what Paul was really saying. First the oldest English version...

> *1 CORINTHIANS 9:20*, {KJV, King James Version} "And unto the Jews I became as a Jew, that I might gain the Jews; to them that are under the law, as under the law, that I might gain them that are under the law..."

That statement alone is damning enough when you use critical thinking to see what he was saying. He uses the term "to gain the Jews… that I might gain them." To the average person, Paul was out to gain members of the Jewish Christologies to join his own. To gain something from them. We know that the system of tithing in Judaism was his real motivation, to bring as many people into his religion as he could persuade from all over the Roman Empire. By forming this religion, as it is in Judaism and many Christian organizations today, converts would be required to give 10% of their income (of any kind) to him and his organization. He wasn't spreading a "Gospel," which is Greek for "good news," just a religious cult. James (Jesus' brother) and the Jerusalem church disagreed with Paul on letting Gentile converts not eat Kosher at the Feast that would take place upon the returning Christ, in spite of what the book of Acts says and other passages in Acts. When we look closely at Paul's own letters with open eyes, it is clear what he was doing.

In the margin of the NKJV (New King James Version) where this verse appears is information that tells us that this verse has another sentence in the middle of the verse according to early manuscripts; "(though not being myself under the law)".

The NRSV (New Revised Standard Version) includes this sentence in the text because Scholars like Ehrman believe it was part of the original text but taken out at some later time by the Orthodox scribes who made the copies that were eventually preserved.

Here is how it is presented in the NRSV. [underlining added for emphasis]

> *1 CORINTHIANS 9:20*, {NRSV} "To the Jews I became as a Jew, in order to win Jews. To those under the law I became as one under the law (though I myself am not under the law) so that I might win those under the law."

Was he being a little deceptive? Yes, absolutely! Is he saying that he was pretending to be a Jew to Jews, even though we know that he really was a Jew from birth, in order to "win those who are under the law" but does not believe that keeping any Jewish law is required by "Christians"? Think about this, what other Laws might he be able to include that he doesn't have to keep? How far could he take it? That's what he is saying. This is what Christianity believes and teaches because this is what Paul vigorously taught in total opposition to what the disciples and James wanted. Decades, if not centuries, of Jewish Christians thereafter are remembered for their opposition to Paul.

As the book of Acts approaches a climax of the conflict between Jesus's own brother James and Apostle Paul there is a definite rift and separation between Paul's following and the original "Church" as well as all Jewish members of Paul's cult. The author of Luke and Acts was the same person and in each case, did not mention that the "James" of the "Jerusalem church" was Jesus's real flesh and blood brother. That alone should open eyes to the truth. Why would they hide the fact that "James, the brother of Jesus" was the very same "James" who controlled what decisions would be made for their sect? When the "Jerusalem Council" met over whether Gentile converts must get circumcised in the 15th chapter of Acts, we see a fabrication near the middle of the chapter and another at the end which hides what happened. In the 13th verse, James announces to the council to "listen to me."

Then he sites a passage from the Old Test book called "Amos" (Acts: Verses 16-17) about God "rebuilding the tabernacle of David… so that the rest of mankind may seek the LORD, Even all the Gentiles who are called by My name, Says the LORD who does all these things." Then in Verse (19) James says, "I judge that we should not trouble those from among the Gentiles who are turning to God, (20) but that we write to them to abstain from things polluted by idols, from sexual immorality, from things strangled, and from blood."

Keep in mind that this was supposedly James the brother of Jesus who took over the leading role of the Jerusalem church immediately after the crucifixion because of his bloodline to King David like Jesus. As Jesus' brother, James would be next in line for the throne of King David if Jesus was acting on the authority of Jewish royalty. However, none of the books in the New Testament mentions James as "the brother of Jesus" except a passing comment by Paul in one of his letters. Then in the next chapter of Acts, Barnabas, a Jewish traveling companion to Paul, wanted someone to travel with them and Paul argued with Barnabas over it. In Verses 37-40, "the contention became so sharp that they parted from one another."

It is convenient that a lie like the one that says James made a decision in favor of Paul and Gentile conversions is followed by a Jewish Barnabas leaving Paul's employment as assistant and his relationship as companion. Here is what really happened. The statement by James was fabricated to look like he was allowing Paul to get what Paul mostly wanted, to a certain extent. History and the bulk of the letters by Paul shows a deepened divide thereafter between himself and all Jews. Even the Epistles that claim to have been written by Paul that are forgeries goes ever deeper into antisemitism. Then, the excuse of Barnabas (a Jew) separating from Paul over wanting someone to join them is a fairly obvious fabrication because the following chapters does not mention anymore Jewish followers of Paul. Scholars know that when the Apostolic Council ended it was not amiable between the Greeks and Romans who followed Paul and the Jerusalem church that was the true followers of Jesus, regardless of what ACTS says. The man who replaced Barnabas was Luke, a Greek, Paul's second assistant, secretary, and companion and the person who the Gospel of Luke was named after but not the actual author of Luke or Acts. The logical conclusion, with evidence in Paul's letters about his growing hatred of Jews in general, is that the council at Antioch with the Jerusalem church representatives led by James did not reach any agreement with Paul on any of Paul's proposals. Church history records that Jewish Christianity who survived the Jewish war formed separate groups for a while before fading away from history and all of them considered Paul a heretic and adversary. It altogether spells out the conclusion that ACTS lied about the agreement between James (the brother of Jesus, which it doesn't mention) and Paul. The Acts of the Apostles were completely about Apostle Paul and promotes an agenda to make Paul look like the true founder of Christianity. In this manor, the Gentile Christology could pretend that Paul's doctrine was endorsed by the leader of the remaining members of the original Jesus movement, one that included at least one of Jesus' brothers, "James the Just."

It is easy to see why so many were attracted to Paul's doctrine of beliefs. No requirement at all from converts. They could now convert to Paul's Christology without changing a thing about themselves. He had insisted on more than just ending circumcision and Kosher laws for converts, he argued in his letters against keeping any statutes of God of any kind for anyone by arguing for faith instead of works. This way they could mold their own beliefs into the Pauline/Gentile doctrine and form their own religious doctrines. The fictitious belief in going to a heaven as a reward or hell as a punishment was not what Paul or Jesus taught but crept into mainstream Christianity much later, followed by Judaism.

Note also, how far could this be taken by Paul's followers? Could "though I myself am not under the law" be used to say that Christians are not "under" anyone's "law" but their own personal values and opinions guiding their thoughts and actions? In other words, can they become delusional in thinking they hold only those laws which they can claim as their own? Is it an excuse to be as rogue and defiant as one wishes? Might they even say that the Holy Spirit instructed them to commit genocide, Old Testament style? Doesn't this licentiousness feed possible destructive delusions? If you have this kind of license to make your own rules you can become and are now very dangerous indeed.

They were talking about how to allow Gentiles into God's kingdom when Jesus returns. In particular, what would be on the menu at the first official Feast of the Tabernacles. Something as simple as that. Paul felt that all rules are off once a Gentile is converted and even the Jewish Christians should no longer be burdened with laws and statutes. That is what ended up happening to Pauline Christology. Wouldn't a Jewish God expect converts to adhere to Jewish instructions on living? The Jewish Christians insisted on making Gentile converts respect and adhere to the vast majority of Jewish customs, traditions, and laws, like "James" did. This was a huge motivation for why the Greeks who wrote the New Testament were writing Jewish followers out of the stories to the extent that Paul's Greco-Roman Christology changed the Jewish names of important people within the movement. And in that first century, "Simon" (Peter) was the most popular name for Jewish men in Judea. This made it associated with Jewishness, which the Orthodoxy/Pauline Christologies that sprang forth wanted to cover up. So, Simon became Peter and Saul became Paul. I suspect he wanted to be called by a Roman and Greek name rather than his thoroughly Jewish name "Saul" and that's why the writers of the Gospels began saying that Simon's name was changed by Jesus to "Peter." Which I am sure the historical Jesus would never have done. When you are told by your God not to even associate with Greeks, and you teach people to obey your God's basic rules, it makes no sense to change the name of someone from a Jewish one to a Greek one.

Think about that. When something does not make sense it is a sign that there is some hidden deception. This is/was a Jewish God of Judaism. They were taught to respect God's name as sacred. They were instructed to stay away from associating with Gentiles. It would be a slap in the face to a fundamentalist Jew to change the names of people in the Jesus movement from Jewish to Greek and/or Roman names.

Paul was the heretic all along, teaching completely different things than what a historical Jesus believed and those who Jesus left behind in Jerusalem. The Gospel writers and Acts tried to cover that up as much as they could through embellishments and fabrications. You can be sure that all the text in the New Testament was written by Greeks with an agenda to manipulate 1st century Christologies away from any Jewishness.

Paul's Christology, which included Gnostic and Marcion influences, transformed into a Greek and Roman mythological religion over the coming three centuries until the emperor of Rome, "Constantine," embraced it for his own political agenda. The picture that Christianity eventually painted was somewhat about the life and death of a man called "Jesus" and by accepting that death as a payment to God for your sins you get easy access on a stairway to heaven.

To believe something "in faith," is to believe something "(Merriam/Webster's) without doubt or question."

Do people realize what that really means? To believe in anything without question or doubt is how you get brainwashed, indoctrinated, and manipulated into giving up your individuality to a cult.

This next passage was the first of the New Testament writings that was written and Paul's first one. It is clear who he blames for the death of Jesus at a time when he was trying to discredit and harass the disciples any way he could…

> *Thessalonians 2:14–161* , "For you, brothers and sisters, became imitators of the churches of God in Christ Jesus in Judea, for you suffered the same things from your own compatriots as they did from the Jews, who killed both the Lord Jesus and the prophets and drove us out; they displease God and oppose everyone by hindering us from speaking to the Gentiles so that

> they may be saved. Thus they have been constantly filling up the measure of their sins; but God's wrath has overtaken them at last." (referring to the Jewish War)

Pontius Pilate... "(*Britannica*) was ordered back to Rome {about 36 CE} to stand trial for cruelty and oppression, particularly on the charge that he had executed men without proper trial." (Only a few years after Jesus)

Does that sound like the Pilate who the Gospels say had sympathy for Jesus and wanted to let him go? Real history tells a different story than the Gospels in many other ways as well, but particularly in that part of the story and the virgin birth nonsense.

> *BART D. EHRMAN*, "Paul, as a person living in the ancient world, knew full well that the reason people sacrificed animals, and in some cultures sacrificed humans, in order to avert the anger of god. God sacrificed his own son, and his son sacrificed his life but not for his own sins, because he's the son of God, he was righteous. He must have died for the sins of others. And so, Paul kept thinking, well, if Jesus was sacrificed for the sins of others, that's how people's sins are removed. Sins are removed by the sacrifice of Jesus. But if sins are removed by the sacrifice of Jesus, what about… I thought within Judaism that people were right with god by following the Law of God. But Paul came to think actually being right with god doesn't come by following the Law. If they came by

> following the Law, there's be no reason for god to sacrifice his son."

And that goes to the heart of Christianity. Paul's doctrine is an excuse to disregard Laws like commandments, kosher laws, statutes, and even the basic standards of morality in our conscience — if we have one. What is stopping them from disregarding common-sense, human decency, and every principle and tenet of acceptable behavior when they try to justify every evil under the sun including genocide and war crimes like during the Holocaust when the Church was favorable toward Hitler?

A conscience?

Repentance, definition: *Wikipedia*, "Repentance is a stage in Christian salvation where the believer turns away from sin."

Notice that it says to "turns away from sin?" Sin is the simple act of breaking God's laws. How many "Christians" understand what sin really is according to the Bible? That includes any of God's statutes as well as His Commandments. How do you get around it? Change the Law! Or in Paul's case, eliminate the requirement of keeping any laws that you don't want to keep.

WIKIPEDIA "Pauline Christianity is a term used to refer to a branch of Early Christianity associated with the beliefs and doctrines espoused by Paul the Apostle through his writings. The term is generally considered a pejorative by some who believe it carries the implication that Christianity as it is known is a corruption of the original teachings of Jesus, as in the doctrine of the Great Apostasy."

53

Excavating Jesus

In *ROMANS 8:14-23* we see where Paul's own writings reveal some adoptionist belief mixed in with Gnostic.

> "(NRSV) For all who are led by the Spirit of God are children of God. For you did not receive a spirit of slavery to fall back into fear, but you received a spirit of adoption. When we cry "Abba! Father!" it is that very spirit bearing witness with our spirit that we are children of God, and if children, then heirs: heirs of God and joint heirs with Christ, if we in fact suffer with him so that we may also be glorified with him. I consider that the suffering of this present time is not worth comparing with the glory about to be revealed to us. For the creation waits with eager for the revealing of the children of God, for the creation was subjected to futility, not of its own will, by the will of the one who subjected it, in hope that the creation itself will be set free from enslavement to decay and will obtain the freedom and the glory of the children of God. We know that the whole creation has been groaning together as it suffers together the pains of labor, and

> not only the creation, but we ourselves, who have the first fruits of the Spirit, groan inwardly while we wait for adoption, the redemption of our bodies."

There may be some Gnostic Christian interpretation from the above passage but the heart of it is based on the original Jewish Christology of the historical Jesus himself. The Jewish Christians of those first centuries made a vow of poverty in order to be humble and serve the needy with a spirit of love for all humanity. We know that from all sources. It is not disputed, but Christianity does not recognize that original Christology as the "correct teaching." They called it something "heretical," even evil, that must be eliminated. And it was.

Some Gnostic Christians in the early centuries believed that the goal/purpose of human beings is to return to the divine realm from whence we came, based on their influences by mostly Egyptian mythologies. These are pagan (non-Jewish) interpretations and beliefs combined to take center-stage in the growth of the emerging Christian religion. They were covering up the original Jewish Christian's purpose and process that an intelligent Designer may have introduced to human history through the historical Jesus — whether by fiat, design, or accident.

> 'The New Testament: A Historical Introduction to the Early Christian Writings', (*Bart D. Ehrman*) page 12, "Books in the ancient world could not be mass produced. They were copied by hand, one page, one sentence, one word, one letter at a time... Since books

> were copied by hand, there was always the possibility that scribes would make mistakes and intentional changes in a book — and every time it was copied. Moreover, when a new copy was copied, the mistakes and changes that the earlier scribe (copyist) made would have been reproduced, while the new scribe would introduce some mistakes and changes of his own...
>
> Unfortunately, we do not have the originals of any of the books of the New Testament, or the first copies, or the copies of the first copies. What we have are copies made much later — In most cases hundreds of years later. How do we know that these copies were changed in the process of reproduction? Because we can compare the thousands of copies that we now have, which range in date from the second to the sixteenth centuries, to see if and how they differ from one another. What is striking is that they differ a lot. In fact, among the over 5000 Greek copies of the New Testament that we have, no two of them are exactly alike in all their details. We don't know how many differences there are among these copies because no one has been able to add them all up."

Of the first three centuries all we have are pieces of parchments. To the logical mind, this brings up an interesting question. If God had intended these writings to be his inspired words for us to live by and find answers in, why not preserve the originals? Why allow so many changes, additions, mistakes in translations, mistakes in interpretation, and even deletions over the course of 2000 years? Why would there be so many differences between denominations and their offshoots? Because everyone would have different opinions anyway since there is no God around to tell us Himself.

> 'Excavating Jesus: Beneath the Stones, Behind the Texts', by *John Dominic Crossan and Johnathon L. Reed.* Page 38, "one strand of Jewish tradition held that God would bring the Gentiles into full community with Jews at that ideal Utopian or eschatological moment in the future when God finally made earth divinely just. Gentiles would then be converted not to Judaism, with male circumcision, for example, but to the God of the entire world. Jews and Gentiles would then feast together with that God on a pure, just, peaceful, and fruitful earth. It was a vision of God's eventual justification and pacification of a violent earth not by the Great Final War at Mount Megiddo (Armageddon) in which evildoers would be finally slaughtered but by the Great Final Banquet [Feast of Tabernacles] on Mount Zion in which evildoers would be finally converted."

Excavating Jesus: Beneath the Stones, Behind the Texts, "Eusebius records that Clement of Alexandria related, "This James, whom the people of old called the Just because of his outstanding virtue, was the first, as the record tells us, to be elected to the episcopal throne of the Jerusalem church." Other epithets are "James the brother of the Lord, surnamed the Just," and "James the Righteous.""

ISAIAH 2:2-4 (NRSVUE)

2. In days to come

the mountain of the Lord's house

shall be established as the highest of the mountains

and shall be raised above the hills;

all the nations shall stream to it.

3 . Many peoples shall come and say,

"Come, let us go up to the mountain of the Lord,

to the house of the God of Jacob,

that he may teach us his ways

and that we may walk in his paths."

For out of Zion shall go forth instruction

and the word of the Lord from Jerusalem.

4. He shall judge between the nations

and shall arbitrate for many peoples;

they shall beat their swords into plowshares

and their spears into pruning hooks;

nation shall not lift up sword against nation;

neither shall they learn war anymore."

ISAIAH 11:9, "They shall not hurt or destroy in all my holy mountain; for the earth shall be full of the knowledge of the LORD."

THAT obviously hasn't happened yet!

Most of the world has become what is called a "plutocracy" or "Oligarchy."

> The "*Merriam-Webster Deluxe Dictionary: Tenth Collegiate Edition*" definition of the word "plutocracy" is this.
>
> "1: government by the wealthy 2: a controlling class of the wealthy"

Here's just a couple of things that Jesus is quoted as saying.

MATTHEW 6:24, (NIV, New International Version) "You cannot serve both God and money."

LUKE 12:15, (NIV) "Watch out! Be on your guard against all kinds of greed; A man's life does not consist in the abundance of his possessions."

How many Christians read those words in their Bibles and then ignore it? All of the ones who are comfortably wealthy. Especially those who think that they are doing God's work by spending their income on producing free literature for their church and counseling members of their church on how their doctrines expect their members to live, think, and vote.

Hypocrisy is the norm in Christian America. They condemn people who are different, who don't conform to their primitive principles of ideology.

MATTHEW 19:21,23-24, (NIV) “Jesus answered, “If you want to be perfect, go, sell your possessions and give to the poor, and you will have treasure in heaven...” “Then Jesus said to his disciples, “I tell you the truth, it is hard for a rich man to enter the kingdom of heaven. Again, I tell you, it is easier for a camel to go through the eye of a needle than for a rich man to enter the kingdom of God.”

Why would that be a requirement for entering “the kingdom of God” or “Heaven” unless the main requirement is “Altruism.”

It’s a deep concern for the lives of others at the expense of your own. It is putting yourself into the shoes of those who struggle to survive more than you might otherwise. Not just members of your family. Not just your friends and acquaintances. Altruism is having more concern for the welfare of others than for oneself. It only seems impossible because we are physical beings who require 4 basic needs; water, shelter, fire, and food. Call them the, “Four Essentials of Survival Science.” The real challenge is to provide all of those essentials for yourselves and share it all openly with those around you who struggle, no matter who they are. Do not judge others! Period!

If we as a species cannot develop an empathy toward all life in general, the greed and apathy that prevails will destroy us. Mark my words. Humanity is in the end-of-days because of our greed, apathy, stupidity, and pious self-righteousness.

“The Mythmaker: Paul and the Invention of Christianity”, by Hyam Maccoby), “The Gospels say quite distinctly that Jesus founded a Church. Why, then, did the Apostles of Jerusalem act as if no church had been formed, and they were still members of the Jewish religious community? This leads to the further puzzling question: if Jesus, as the Gospels say, chose Peter as the leader of the Church, why were the Nazarenes, after Jesus' death, led not by Peter, but by James… a person who is not even mentioned in the Gospels as a follower of Jesus in his lifetime? This is the kind of contradiction that, if logically considered, can lead us to the true picture of the history of Jesus' movement in Jerusalem, as opposed to the picture which the later Church wished to propagate."

54

It Hasn't Happened Yet

We are frequently seeing what religion does to people in the Middle East. They did not adopt the teachings of Jesus, "turn the other cheek," or "pray for your enemies." However, modern Christianity does not live by Jesus' teachings either. Now we have a Christian theocratic movement in America trying to destroy democracy to replace it with evangelical theocratic dictatorship with them deciding which parts of the Biblical rules to enforce on us and which to ignore. They are the ones who think that banning books is a good idea. Sure, let's limit everyone, especially young minds, to only reading what we choose for them. It's absurd, especially when a book by Albert Einstein has been one of the banned books.

The Old Testament orders followers to pay a "Tithe," ten percent of their income, to their religious leaders. It was originally intended to mean a percentage of food, tools, and other necessities in order to make sure that everyone's needs were being addressed. Now it pays for a tax-free industry of property and building owners to "worship," produce their religious/political propaganda with literature and video, and provide the lavish lifestyles of the most apathetic, self-centered, and egotistical, sociopaths in humanity today. Their delusion convinces them that "God" gives them their comfort and wealth to live by their own rules which they perceive are divinely ordained for everyone to follow — or, in other cases, everything is given to them by their perception of a god. That has always been a dangerous thing for everyone else to have to deal with.

> *ZECHARIAH 14:16*, "And it shall come to pass that everyone who is left of all the nations which came against Jerusalem shall go up from year to year to worship the King, the LORD of hosts, and keep the Feast of Tabernacles."

Has that happened yet? No. And they never really decided on what kind of food would be offered at the Feast, Kosher, or non-Kosher.

The original church of Jesus and James said it must be Kosher when Paul and his Greek and Roman friends wanted to make it non-Kosher.

> *ZECHARIAH 14:1,4,5b,6-9,11,* (NKJV) "Behold, the day of the LORD is coming... And in that day His feet will stand on the Mount of Olives... Thus the LORD my God will come, And all the saints with You (translated, "you", or "Him")... It shall come to pass in that day That there will be no light... It shall be one day Which is known to the LORD — Neither day nor night. But at evening time it shall happen That it will be light. And in that day, it shall be That living waters shall flow from Jerusalem... And the LORD shall be

> King over all the earth. In that day it shall be — "The LORD is one," And His name one... The people shall dwell in it; And no longer shall there be utter destruction, But Jerusalem shall be safely inhabited."

THAT hasn't happened yet. That would be when the proverbial Day of the LORD is prophesied to come. No one has been "Saved" from anything yet.

> *ECCLESIASTES 1:13-17*, (NIV) "I applied my mind to study and to explore by wisdom all that is done under the heavens. What a heavy burden God has laid on mankind! I have seen all the things that are done under the sun; all of them are meaningless, chasing after the wind. What is crooked cannot be straightened; what is lacking cannot be counted. I said to myself, 'Look, I have increased in wisdom more than anyone who has ruled over Jerusalem before me; I have experienced much of wisdom and knowledge.' Then I applied myself to the understanding of wisdom, and also of madness and folly, but I learned that this, too, is a chasing after the wind."

"Alive and Well with Wild Foods" by a Peter Ragnar. Some statements related to this subject caught my eye on pages 70-71 about an herb called "Soma."

Alive and Well with Wild Foods, "The mysterious Soma has its origin in a most unlikely celebration: the Eucharist, or Holy Communion. The celebration of the bread and wine goes back much further than Christianity. For instance, Cicero, who was born in 106 BC (That's 106 years before the alleged birth of

Christ), stated, “How can a man be so stupid as to imagine that that which he eats to be a god?” The great Roman orator was criticizing a practice that was far older than even he suspected. The sacrament of the bread and wine can be traced to Melchizedek of Bible fame, as well as to ancient Tibet and Nepal... One of the first Christian missionaries was a French man, Andrada La Crozius, who wrote about the sacrament in a writing called Anacalpsis. This was nothing new to the ancient world.

Even the Greek sage Pythagoras, who awed young Greek wrestlers at one hundred years of age, practiced this ancient rite. He was said to have learned it from Egypt. Now keep in mind, he was born 570 years before the Christian era... According to Mr. Baring-Gould, in his book Original Religious Belief, “Among the ancient Hindus, Soma was a chief deity; he is called 'The Giver of life and health, the Protector, he who is the guide to immortality.' He became incarnate [definition, “endowed with a human body; personified.”] among men, was taken by them and slain, and brayed in a mortar. But he rose in flame to heaven, to be the 'benefactor of the world,' and the mediator between God and Man. Through communion with him in his sacrifice, man (who partook of this God), has an assurance of immortality, for by that sacrament he obtains union with his divinity.”

Wow, that should put to rest how the Last Supper ritual and the Holy Communion came to exist in Christianity. It was borrowed from ancient mythologies from a variety of places and times. But wait, there’s more revealing information there.

55

Quantum Entanglement

Are you curious about the world, about life, and about what we are with an open-minded but skeptical approach to filter what you see, read, and hear through logic and common sense? You should. You started out life that way, but something may have changed you and that might be "indoctrination."

per *Wikipedia*: "the process of teaching a person or group to accept a set of beliefs uncritically." "(example) prisoners are subjected to brainwashing, indoctrination, and punishment"

It didn't change me. Nothing about my religious upbringing made much sense to me but there didn't seem to be an alternative of logic and reason until after many years of intense research.

The universe is not a "Matrix", like the movie series, but close.

Quantum mechanics reveals a phenomenon. Subatomic particles, the parts of the atom, like electrons, protons, and quarks, (including photons) do not possess any mass, the 3 dimensions (plus time) that we associate with our reality.

"Photons", per *Google*, "a particle representing a quantum of light or other electromagnetic radiation. A photon carries energy proportional to the radiation frequency but has zero mass."

Photons are superimposed (placed or laid over something else, typically so that both things are still evident, Oxford). And they are "entangled" (became twisted together). Their behavior and reactions are spontaneous with each other, acting together as a whole regardless of distances apart. This is why they are called "entangled" and "interconnected." The universe, at the subatomic particle level, displays "interconnectedness." That was one reason why Stephen Hawking concluded that the universe is a "living organism." All parts of the atom act simultaneously throughout the universe as if it was all one entity and doesn't exhibit any inherent properties; neither solid, liquid, or gaseous — and they react differently when being observed than not being observed. They have no width, length, or height with maybe the exception of the nucleus of the atom.

Does the nucleus contain the mass of an atom?

DOE Explains...Nuclei (*Department of Energy*)

> "The nucleus makes up much less than 01% of the volume of the atom, but typically contains more than 99.9% of the mass of the atom. The chemical

> properties of a substance are determined by the negatively charged electrons enshrouding the nucleus."

So, what is that other 99.9% of the rest of the atom? Nothing. It is an empty space. It is not solid, liquid, or gaseous.

The leading science of theoretical physics and Quantum mechanics support this. Albert Einstein called "entanglement," "spooky action at a distance."

"*WHAT IS ENTANGLEMENT*"(Entanglement – EpiQC)

> "Entanglement describes the relationship that exists between two or more particles that interact in a way that makes it impossible to describe each particle independently. Rather, measurements of the particles reveal correlations, and as such the particles must always be described as a quantum state of the whole system."
>
> (*scienceexchange@caltech*)" Entanglement is at the heart of quantum physics and future quantum technologies. Like other aspects of quantum science, the phenomenon of entanglement reveals itself at very tiny, subatomic scales. When two particles, such as a pair of photons or electrons, become entangled, they remain connected even when separated by vast distances. In the same way that a ballet or tango emerges from individual dancers, entanglement arises from the connection between particles. It is what scientists call an emergent property."

In the physics of Quantum mechanical terms, in any fluctuation of these "virtual particles" the quantum field is being determined by the "uncertainty principle," (not a random principle, just uncertain) which is the source of everything in existence within our universe and in every consciously self-aware being — each one projecting their own experiences and appearance within an interconnected and biologically living universal environment. Our spacetime causality — energy information and matter — are irreducible in its complexity (cannot change randomly) which reacts to being watched by consciousness. It may be that we are the only conscious life-forms in this specific galaxy.

As Professor Brian Cox says, "We give this galaxy meaning by being conscious of it. If we mess that up, by going extinct, we may be eliminating meaning forever from this galaxy. (paraphrased)"

He also recently said that out of the 200 to 400 billion stars in this galaxy, we know that there are potentially a million Earth-like planets. However, astrobiologists would say that it is only a potential and very unlikely that many, if hardly any, have intelligent physical life similar to us. Which must be the reason that Professor Cox said that we might be the only sapient, sentient, and self-aware life-form (with a conscious mind) in this galaxy.

When the Christian world realizes their mistaken belief in the immortal soul from pagan origins and embrace the resurrection doctrine they will have the opportunity to consider the Quantum holographic universe theory of a Great

White Throne Judgment (Revelation 20:11) being scientifically possible someday by the elusive mind behind the design, then, at long last, a real connection between science and religion can be accomplished. However, Christianity will also need to realize the mistake of the ludicrous prosperity gospel and sacrificing a human to a god.

Bart D Ehrman, *UNVEILING THE PROSPERITY GOSPEL,*

> "The prosperity gospel is the idea that God wants you to give to the church and in return God will enrich you. You give of what you have, and God will then enrich you, and God wants you to be rich. That's the exact opposite of Jesus' teaching! Jesus' idea of life is that it you want to follow him, you have to have his mindset, the desire to help others. Then you can "take up your cross" and "follow me." The ORIGINAL Christian teaching was that Jesus shows us how to live. He gives his life for others. He doesn't become rich and famous like a ruling king who has everything at his disposal. Jesus thought you have to give up your life for the sake of the coming kingdom."

Richard Dawkins, "The God of the Old Testament is arguably the most unpleasant character in all fiction: Jealous and proud of it, a petty unjust-unforgiving control freak; a vindictive, bloodthirsty ethnic cleanser; a misogynistic, homophobic, racist — infanticidal, genocidal, pestilential, megalomaniacal, sadomasochistic, sadistic, capriciously, malevolent bully."

If you think you understand the theoretical physics of Quantum Mechanics, you don't understand it. Hawking's last words on the subject included seeing the universe as more of a biological living organism than any definitive mathematical equation or mechanical construct. Einstein called the entanglement of Quantum Mechanics, "spooky action." Hawking also proclaimed there will never be a unifying theory between General Relativity and String Theory. They don't know what Dark Energy or Dark Matter is, but they are sure it does exist as the answer to Einstein's hypothesis called "Cosmological Constant", which the world of physics said didn't exist at the time but has since accepted.

The science community is sometimes as reluctant to accept being wrong about something important as religion is. That is why they don't want to consider deist designed evolution a theory. Everybody wants to think they are right about everything. That may be the greatest impediment to developing a benevolent nature as anything there is.

56

Vegetarians in the Garden of Eden

It can be assumed that by the second century, the Jewish Christians were divided into different forms of mutations from some of their original beliefs. However, in the Jewish tradition of religion you tried to keep somewhat close to the original beliefs no matter how far outside of the traditional doctrine you had ventured.

So far, I have learned that some of the Jewish Christians were called "Ebionites" by Orthodox church historians. There were other groups of Jewish Christians, and their teachings were always in favor of the view that everyone, Jews and Gentiles, will be required to understand and keep God's laws and statutes when the Day of the Lord happens in accordance with how Jesus and James from their messianic apocalyptic Jewish sect would have understood it.

These Jewish-Christians refrained from eating meat because most meat was procured through the religious act of sacrificing the animal to a god. In this abstention from eating dead animal flesh they may have developed a real empathy for animals. Empathy for all life is exactly what we need to evolve more of to qualify for the process of furthering the evolution of humans.

When I tried adopting the different kinds of vegetarianism, I also found myself being at least more respectful toward animals in general. Developing increasing amounts of empathy and concern for all life and its place in our ecosystem and universe is in reality a requirement to fulfill our purpose, if it is our purpose, to become the literal offspring of the Designer, or a hybrid life-form, or simply a more advanced version of humanity. A further evolved humankind.

There were plenty of religions that included sacrifices to gods, not just Judaism. However, to get people to stop killing animals I can see the possibility that the universe would not want us killing our animal kingdom for food when we have "the herb of the field." We are all beasts, we were all made from the same atoms as everything else. WE ARE LIFE. We are the universe. Why kill any living thing at all if we don't have to? It appears that a vegetarian diet is about the most healthy kind if done in moderation.

> (*newsinhealth.nih.gov*) "Vegetarians tend to weigh less than meat-eaters, and to have lower cancer rates. "Evidence also suggests that a vegetarian diet is associated with a lower risk of death from certain heart diseases, and that those who follow a vegetarian diet tend to have lower LDL ["bad"] cholesterol levels," says Engel."

GENESIS 1:29-31, (NRSV) "God said, "See, I have given you every plant yielding seed that is upon the face of all the earth and every tree with seed in its fruit; you shall have them for food. And to every beast of the earth and to every bird of the air and to everything that creeps on the earth, everything that has the breath of life, I have given every green plant for food." And it was so. God saw everything that he had made, and indeed, it was very good. And there was evening and there was morning, the sixth day."

(*Rutledge*)"In Greek mythology, it was said humans were sacrificed to Zeus at the king of the gods' birthplace on top of Mount Lykaion. In one tale, the king Lycaon turns into a wolf after he sacrifices a baby on the altar there. In another tale, a boy is sacrificed with animals, and all the meat is cooked together."

(*Wikipedia*) "Where did human sacrifice come from?

Funeral human sacrifice was widely practiced in the ancient Chinese state of Qin. According to the Records of the Grand Historian by Han dynasty historian Sima Qian, the practice was started by Duke Wu, the tenth ruler of Qin, who had 66 people buried with him in 678 BCE."

Examples of human sacrifice in historical religion included (*thecollector.com*),

> "The European cultures of antiquity were also no stranger to the ultimate religious expression of violence. The Greeks, Romans, Celts, and Germanic

> peoples all practiced ritual human sacrifice to a certain degree.
>
> The ancient Greeks practiced "pharmakos," whereby cripples, criminals, or slaves would be thrown off a cliff to ward off disaster. This process disappeared in later antiquity as the Greeks moved towards a more civilized society. In Roman society, human sacrifice was rare but better documented. Like the Greeks, human sacrifice was seen as a remnant of a more uncivilized age, and as Rome progressed, the victim was replaced with symbolic items such as dolls or statuettes."

A plant specific diet is important to know about, moving forward.

We once believed that cancer was strictly a genetic malfunction and primarily caused by hereditary family history. However, cancer runs rampant in places where the air, water, and food are toxic from being exposed to many carcinogenic compounds in many things. In our non-organic food products we find chemical pesticides, herbicides, and fertilizers or meat animals which are fed with growth hormones, antibiotics, and other substances. We see increasing amounts of water everywhere being contaminated with the worst toxins known to man, even radiation.

Defective genes which run in a bloodline make certain people more prone to developing a cancer, but it is like loading a gun — it takes an environment of bad food, water, and air combining together with carcinogenic compounds that pulls the trigger. All processed meat are a class 1 carcinogen. The best diet to take is all plant based with no oils, dairy, surgery things, and anything with a face. My siblings should heed the warning of our family's fate. Cancer and degenerative disease were at least partly caused by the diet we were raised on.

When we eat the wrong foods, we can trigger the genes that cause cancer and/or spark other ailments and diseases by making us vulnerable. Our genes are affected by the environments we live in. Good food (particularly high in antioxidants) actually protects us in so many ways that we haven't discovered them all.

With over 8 billion people in the world today, it's harder to feed them all and cost is always a factor. So, mass produced food products are now chemically and genetically altered. But not for the nutrition and health of the masses; for flavor, cost, quantity and especially in our world, profit. How will humankind feed themselves once the Sixth Great Mass Extinction ramps up depends on what they learn about healthy food and wild plants that have edible and medicinal qualities as well as other useful purposes. It was for this reason that I co-authored the book in 2020 with survival expert Mykel Hawke called, "*FORAGING FOR SURVIVAL: Edible Wild Plants of North America.*"

The Creation Story in Genesis was interpreted to imply that we were made separately from the animals. According to Genesis, we were the last things created, made on the "sixth day." This has been interpreted literally to being a supernatural event. The Holy Bible implies that we are "mortal" and "flesh" and breathe the same "breath of life" as the animals. We all "go" to the same place when we die, "the dust of the earth from which we were created" and nothing more, at least for now.

There is overwhelming proof in the Holy Bible that "God" wanted humanity and animals to all be vegetarian. According to Genesis and beyond, we were intended to live in peace and cooperation with animals as well as each other. Crazy? To this crazy mixed up world, yes! However...

Eating dead animal flesh was necessary for the survival and growth of early humans, the "hunter/gatherers." A migration of today's modern species of humanity about 75,000 years ago came out of South Africa heading north and northeast. These human beings followed large herds of animals while hunting them for food, shelter material, tools, and clothing. Much earlier human species, over 800,000 years ago, appeared to have a definite increase in their intelligence once meat became a large part of their diets. Learning how to hunt animals the most effectively forced them to use their minds more efficiently while the added meat-proteins developed their brains much further than before.

The Holy Bible says that God told the Hebrews not to eat certain animals, calling them "unclean." Other animals were "clean" meat, and therefor edible to man in God's eyes. A notable characteristic of most "clean" animals is that they are vegetarian themselves, herbivores. A common characteristic of clean fish is that they have both fins and scales. The most common characteristic of unclean sea creatures is that they are scavengers or bottom-feeders, like all shellfish. The logic behind this is actually quite remarkable for such a primitive people, given the truth as we know it now of how safe or unsafe certain seafood can be.

However, why and how this "God" decided to allow his "chosen" to eat meat at all is interesting when a close look at the first chapters of Genesis indicates conclusively that "God" had commanded that all animals and humans be vegetarian in the "garden of Eden."

To begin with, The Bible makes the claim that "God" has given this planet to humankind.

PSALM 115: 16, "The heavens are the LORD'S heavens, but the earth He has given to the sons of men."

(Quotations from the Bible for this chapter comes from the New King James Version, NKJV unless noted otherwise. Underlining and brackets added in places for emphasis.)

GENESIS 1:28-31, "God said to them, "Be fruitful and multiply, and fill the earth and subdue it; and have dominion over the fish of the sea and over the birds and over every living thing that moves upon the earth." And God said, "Behold, I have given you every plant yielding seed which is upon the face of the earth, and every tree with seed in its fruit; You shall have them for food. And to every beast of the earth, everything that has the breath of life [oxygen breathing life] I have given every green plant for food."

And it was so. And God saw everything that he had made, and behold, it was very good."

A "beast" in the Bible is an animal.

This is plain English we can all understand without spinning, twisting and turning a phrase here and there to interpret it some other way. This "God" character is saying that he had provided "you every plant yielding seed" and "every tree with seed in its fruit" for "food" and to every animal he has given "every green plant for food." This God was basically commanding that Adam and Eve and all animals to be vegetarian while in his garden. How can one NOT interpret it that way?

In reality, a Designer of some kind designed the universe to be biofriendly. Our little corner of the universe produced a perfect human habitation by purely natural causes and processes certainly not by some supernatural fiat or by divine guidance and direction and yet, the ancient people believed it was actually a supernatural instantaneous "creation" as does many gullible religious people even today.

After Adam and Eve broke God's command concerning the forbidden fruit he told Adam,

GENESIS 3:17-19, "Cursed is the ground because of you. [Also translated "for your sake"] In toil you shall eat of if all the days of your life... and you shall eat the plants of the field. In the sweat of your face you shall eat bread till you return to the ground." VERSE 23, "therefore the LORD God sent him forth from the garden of Eden, to till the ground from which he was taken."

As a consequence of Adam's removal from the garden, where fruits and vegetables were abundant, he would now struggle to "till the ground" in the "sweat of your face." However, did Adam and Eve and the animals continue to be vegetarian as God had originally instructed? Didn't they go back to being carnivorous?

> *GENESIS 6:11-13*, "Now the earth was corrupt in God's sight, and the earth was filled with violence… for all flesh had corrupted their way upon the earth" And God said to Noah, "I have determined to make an end of all flesh; for the earth is filled with violence through them."

According to scholars, the proper translation for the word "earth" may actually be "land." The text would not necessarily include the entire planet, just the land of Eden and surrounding area. These are writings that date way back before we understood that we live on a planet. We did not know where the sun went at night. We had very little clue about anything until recent centuries.

In *GENESIS 2:19*, God brought "every living creature" to Adam "to see what he would call them; and whatever the man called every living creature, that was its name."

If God sent every animal to Adam, you can be sure that they did not try to harm each other or harm Adam. If we are supposed to believe this story, surely God's influence of selflessness and benevolence was canceling out the animal instincts of humans and animals. In another example, even though the story is very fictitious, the animals aboard Noah's

Ark would not have harmed Noah, his family, or each other if it was divinely ordained.

> *GENESIS 6:17*, "... I will bring a flood of waters upon the earth, to destroy all flesh in which is the breath of life [oxygen] from under heaven [the sky]; everything that is on the earth [land] will die."

The earth was filled with violence because all flesh had corrupted their way in God's sight. Not only was the descendants of Adam and Eve hunting animals and competing against each other in the land of Eden, but animals were killing each other and eating people. This would be contrary to God's way as instructed earlier. When this God had been in control of the garden of Eden where he planted those two special trees, the "tree of life" and the "tree of the knowledge of good and evil", "He" could control the behavior of the animals. After the expulsion of Adam and Eve from that garden, this God had relinquished control back to the natural forces of this material world and its animal instincts. I know it is a stretch of the imagination but if religion is going to present fairy tales as history, at least these interpretations that I present for you here can make a resemblance of common sense and logic.

GENESIS 7:2, Noah was "the first tiller of the soil."

Among the many biblical prophesies in the Old Testament "book" called "Isaiah" about a coming apocalyptic "kingdom of God" to the earth there is a vivid description of people and animals all living peacefully together.

> *ISAIAH 11:6-7*, "The wolf shall dwell with the lamb, and the leopard shall lie down with the kid [young goat], and the calf and the lion and the fatling together, and a little child shall lead them. The cow and the bear shall feed; their young shall lie down together; and the lion shall eat straw like the ox."

If this describes a picture of when a "messiah" rules the world or after the Great White Throne Judgment is over, it resembles those first days in the garden of Eden when all animals and those two human inhabitants were commanded, and somehow influenced, to be vegetarian... if the story has any truth to it.

Religion might insist that the verses above about animals living peacefully together is just mere symbolism, that it only symbolizes peoples and nations living in peace with other peoples and nations.

Consider, if the Designer of the universe could influence all people to live this way with each other someday, would it be such a stretch of the imagination to include "all flesh in which is the breath of life"?

Our own animal instincts of self-preservation influences us to be self-centered, that's just part of being this material living thing. What every human has to choose between sooner or later is whether to care about others or only care about oneself and the extensions of oneself. There is no God involved in our character choices even if a "God" of some kind exists. If we are a design, we were designed to evolve benevolence in spite of our animal instinct to be self-centered. The reason for our suffering and our inability to create a peaceful world is not the

DESIGNER'S fault or some magical, imaginary adversary, it's our own. Once/If humanity can get past the temptation to be self-centered and hateful toward others and become powerfully empathetic toward all living things (particularly other humans) all of the problems that plague humankind will be solved. You can find many examples of Jesus talking about how we should treat each other. If we are to evolve into a benevolent state of mind we must be ready for whatever is the actual purpose behind the design.

57

"Neither Shall They Learn War Anymore"

ISAIAH 2:4 & MICAH 4:3, "He shall judge between the nations, and shall decide for many peoples; and they shall beat their swords into plowshares, and their spears into pruning hooks; nation shall not lift up sword against nation, neither shall they learn war anymore."

IMAGINE THAT...

All weapons shall be banished from the entire world with this DESIGNER'S power and influence over "all flesh." Today's weapons will become tomorrow's farming tools.

ISAIAH 51:3, "For the LORD will comfort Zion; he will comfort her waste places, and will make her wilderness like Eden, her desert like the garden of the LORD."

ISAIAH 11:8-9, "The nursing child shall play by the cobra's hole, and the weaned child shall put his hand in the viper's den. They shall not hurt or destroy in all My holy mountain, for the earth shall be full of the knowledge of the LORD as the waters cover the sea."

ISAIAH 35:5-8, "Then the eyes of the blind shall be opened, and the ears of the deaf shall be unstopped. Then the lame shall leap like a deer, and the tongue of the dumb sing. For the waters shall burst forth in the wilderness, and streams in the desert. The parched ground shall become a pool, and the thirsty land springs of water; In the habitation of jackals where each lay, there shall grass with reeds and rushes."

Imagine a rhinoceros plowing your field with his horn? Cows can fertilize the soil right where you need them to. Imagine large animals like horses, bears and elephants clearing the land for crops, working construction of a home or barn, or providing transportation in complete cooperation with you without any training. How about goats eating garbage that you cannot compost and the most useless of weeds without touching your crops and herbs? What if you could have giraffes and monkeys picking and packing fruits and vegetables alongside human beings? Is it such a stretch of imagination, that the Designer of the universe could/would build our planet like that if IT, He, She, or whatever, suddenly wanted to as part of a plan to complete the specific purpose for creating this universe? Imagine having every living thing on the Earth living in complete peace and cooperating unselfishly in harmony with each other? Now imagine making it rain where, when, and how much? Is it such a stretch of the imagination to suppose that the Designer of all this could control all of the forces of nature on our planet, in our galaxy, and throughout the universe if and when IT chose to?

There is neither evidence nor practicality of this Designer guiding anything that has happened on this planet so far. A plan of this magnitude would require keeping us all in the dark until the right time comes.

It took billions of years to bring animals and us into existence here on this one planet. Having respect for the Designer's handy-work and for other living flesh reflects that Designer's own nature as a possibly omnipotent, benevolent being and may be a requirement from us if we want to be part of the overall plan and purpose.

Religion's interpretation of the first chapter of Genesis has led them to assume that we did not evolve from animals. They would absolutely deny that there is any biblical evidence that would indicate our animal ancestry. They see themselves as completely separate from animals and superior.

Our bodies were designed to operate efficiently as long as the right materials are consumed. As we reach old age, this material body's ability to regenerate cells begins to falter and eventually fail. Some diseases, like cancer, occur when the regeneration process becomes unbalanced. The one most crucial factors to living a long healthy life is to consume the materials that we were designed to use the most efficiently. A plant-based diet has proven to be the most beneficial.

Animals instinctively try to consume the best materials that their bodies were designed for as natural selection evolved them. However, if the best food for them is not available they will consume any alternative food sources that they can find. The alternative food sources will not keep them healthy, sometimes even making them sick, but they basically know what is good for them and what isn't. Sometimes a brightly colored plant is a warning sign not to eat any of it. It often turns out to be a warning to all herbivores and omnivores. The same with brightly colored insects.

Humans are quite different. While we have basically the knowledge of what we need to eat the most, we prefer to consume the materials which we enjoy the most, usually substances that have been enhanced with salt, oil, and sweets. We ignore that which is best for us in favor of what tastes the best to us. It amazes me that people are brainwashed by this self-deception. It's the young child in us that doesn't have the wisdom to understand why we eat. With my health problems in my youth, I knew early on, without being told by anyone (that I can remember), to find out what I need to eat and adjust my diet to suit that need. My family was not very supportive because of the Standard American Diet (SAD) that everyone expected from everyone else.

The Plant kingdom is crucial to the habitability of our planet. If all life evolved from a common ancestor, it means that we were once plants. I imagine myself a weed who must struggle alone among a variety of grasses, herbacious plants, vines, trees, flowering beauties, and shrubs with my fellow weeds keeping their distance from me. All members of the plant kingdom are more aware of their surroundings than you probably think. Researchers think they are self-aware but not sapient or sentient. Plants are living things, biological entities in a sense, and a very important member of Earth's living family. They absorb CO2 from the atmosphere and release oxygen. That's the opposite of what oxygen breathing life does. They provide nourishment to many forms of life. So, why do humans take them for granted. You decapitate grass, pull weeds, and burn what you don't want. You poison the ground where they live, which also happens to be where you live. There have been many people lost in the woods who starved to death near edible wild plants, roots, tubers, berries, greens, nuts, etc. That's right! People have starved to death where there was food that they did not know about because of a society that lost that knowledge to commerce by ignoring it. That is the very definition of ignorance.

If you don't see where I am going with this your chances of seeing a life of any kind after death is nearly nil. You are underdeveloped humanoid beings who will cease to exist when you die because the Impersonal Universal Consciousness decided that it would be in vain to resurrect people who can't relate to Altruism. It is much more than just caring about the welfare of other human beings. We can't risk having some dumb-ass chopping down the tree of life.

We like to think of ourselves as the only form of life that matters at all to us. Not others, just our own species and our own lives. Survival instinct focuses on the needs of the one, not the many. That will be our foolish mistake in the long-term forecast of coming mass-extinctions. Your Babbling Bible says that you have "dominion."

WEBSTER'S, "Dominion", "domain: Supreme authority: absolute ownership."

GENESIS 1:26, (NKJV) "Then God said, 'Let Us make man in Our image, according to Our likeness; Let him have dominion over the fish of the sea, over the birds of the air, and over the cattle, [so, seafood, poultry, and red meat. This god thinks with a human stomach] over all the earth and over every creeping thing that creeps on the earth according to its kind.'"

Now ask yourself, who are the "Us" in "Let Us make man"? It was the gods, plural, of primitive ancient belief systems. They did not believe that just one god existed, that came thousands of years later when it no longer made sense to our slowly evolving intelligence. Now we are finding out just how important and connected all life is in this collective of earthly living things. We didn't just evolve from animals, we are dependent on all life in order to exist. Living things feed on other living things to regenerate our bodies because that is the way the universe made us. The fictitious gods did not make us. It isn't magic, like your ancestors believed it was. It wasn't gods being gods.

It is physics and biology working/functioning together as it was all designed to do by a MIND of some kind since the beginning of time. But with 9-billion people here now, we are foolishly exploiting our furry and feathered friends {our long-lost ancestors} and their resources to appease the greed of capitalist billionaires who are killing off all living things wholesale until there will be nothing left. It is one thing to be given "dominion" over everything by fictitious gods that we imagined from your ignorance but now we must face the consequences. It is the consequences of having dominion without wisdom. Like Adam and Eve in the parable had to face the consequences of the foolish act of eating a toxic fruit when they had access to every kind of tree that had edible fruit on it.

I am a fan of other species of life. I grow plants where there is none with focus on the edible/useful plants that our foolish society treats like trash. Like fruit trees and wild herbs are called, "trash trees" and "weeds" by a shortsighted general public. It doesn't matter that a wild yam vine can produce a 100 pound yam that can feed starving people, if the average spoiled stupid American doesn't appreciate it's resource value and aesthetic beauty, they kill it without a second thought. They sometimes even leave the most invasive and poisonous plants alone as they are killing useful plants. In my mind it is beyond stupidity. It's mass insanity!

Why does society ignore the needs of countless plants, animals, and people? The self is the only thing important to humanity, not any other. It is the excuse of a society who believes in a "god" to reject thoughts of caring about other living things. If we continue this kind of delusion there will be nothing left for any living thing to feed on. IMHO that is now inevitable.

I am agnostic, as I'm sure you know by now. From a very special (and very large book) I have this information about being "agnostic."

"Encyclopedia of Word and Phrase Origins — Robert Hendrickson (QPB, second edition, Facts On File, 2004)
"AGNOSTIC. "Nineteenth-century British scientist Tomas H. Huxley coined the word 'agnostic,' one who believes there is no proof of God's existence but does not doubt the possibility

that God exists. The anonymous rhyme, "I do not know, it may be so," sums up the position. Huxley, who described himself as "a man without a rag or label to cover himself" coined his label from the Greek a, "without, not," and gnostic, which is related to the Greek gnosis, knowledge."

We should never stop learning stuff. My early years of reading novels and fiction did no good at enhancing my general knowledge of things. Only when I focused on non-fiction is when it felt like my brain grew in size and capacity. Like the "guides" on survival, religion, science, and nutrition — all the subjects with critical information to know for one whose luck blew out early. What have you learned now? Barbecue? Team sports? Favorite restaurants? You are never to old to learn some really important stuff, maybe just too lazy and a little bit arrogant.

The bottom line of our existence, if there is one, is to develop a different kind of nature than the one we inherited from the animals. I have described it with the word "Altruism." The overwhelming concern for others, including other life-forms, at the expense of one's own stability. It is the hardest thing for many people to even imagine. It goes against everything that any material based life-form would naturally have. Here's why it is so important. Adam and Eve reacted like self-centered material beings and took a fruit from a tree that seemed harmless, looked like it would taste good, but mostly because they were told not to and that it might give them a form of wisdom that they didn't already have. But by losing access to the tree of life, they found out that it failed to produce any of that and when they died, that was the end of them. Think of

Altruism as a pathway to the tree of life. It doesn't exist in our world now but we are evolving in that direction.

58

The Gospel of Q

If you want to know what a real Jesus said out of what is quoted him saying, if there was such a person, you might look at the earliest followers and what they may have been saying in oral traditions that someone wrote down and ended up in the Gospels. If fact, scholars have determined that much of what was quoted in the Gospels probably was real saying of Jesus, but long sermons or word for word quotes were less likely spoken by a real Jesus. So, the question remains, what can we know? Is there an outline of the differences between "Jesus": the messianic leader of a Jewish cult, a Teacher of Jewish customs and wisdom, a man of the world (no particular political or religious ambitions), or a humanitarian with altruism. The jury is still out to most scholars, but we know now that the Synoptic Gospels had a source for their versions that came from the earliest followers of Jesus and that points in the direction of the humanitarian Jesus.

New Testament critical scholars have known for quite a while now that the Synoptic gospels of Matthew, Mark, and Luke were using some very early "sayings" of Jesus from an unknown source, called the "Q", within the pages of their versions of the life, death, and alleged resurrection of their "narrative gospels."

A close look at the work of Burton L. Mack from his 1993 book, "The Lost Gospel: The Book of Q & Christian Origins" (*HarperSanFrancisco*). This writer describes the Gospels that were included in the Christian Bible as "narrative gospels" and the Q as the "myth of origin."

> *page 7*, "Q's purpose in attributing sayings of Jesus and its careful design can be seen as the creation of a highly crafted and profoundly effective myth of origin. This myth of origin claimed epic and divine authority for Jesus as a founder figure without the need to entertain mythological notions of a crucified and resurrected messiah." Page 9, "Traditional systems of honor based on power, wealth, and place in hierarchical social structures were called into question, as were codes of ritual purity, taboos on intercourse with people of different ethnic roots, and taxation economics. People were encouraged to free themselves from traditional social constraints and think of themselves as belonging to a larger human family. As Q puts it, "If you embrace only your brothers, what more are you doing than others?"
>
> At first no one was in charge of the groups that formed around such Q teachings. Conversation and mutual support were enough to encourage an individual to act "naturally," as if the "normal expectations of acquiescence to social conventions did not apply."

> "As groups formed in support of like-minded individuals, however, loyalty to the Jesus movement strengthened, a social vision for human well-being was generated within the group, and social codes for the movement had to be agreed upon. Why not ask when in need and share what one had when asked, they wondered? Eventually, therefore, the Jesus movement took the form of small groups meeting together as extended families in the heady pursuit of what they called "God's kingdom.""

Unfortunately, "Despite these agreements, however, every group went its own way and drew different conclusions about what to think and do."

> *Page 10*, "The narrative gospels can no longer be viewed as the trustworthy accounts of unique and stupendous historical events at the foundation of the Christian faith. The gospels must now be seen as the result of early Christian myth-making. Q forces the issue, for it documents the earlier history that does not agree with the narrative gospel accounts."

As we suspected, the Gospels are a majority of myths that grew from the sayings of Q and embellished with miracles of every kind. According to the timeline presented by Mr. Mack, the Gospel of Mark was written around 80CE in Southern Syria. That's 50 years after the crucifixion event and well after the start of the Roman/Jewish war. The first of the Q sayings was written around 50CE, about when Paul's first letter was written. By the time you get to the most influential narrative gospel, Luke/Acts, the years is 110CE, clearly 90 years after the death of Jesus and when 2 Peter was written in 140CE, one can't help but see the controversy of saying that these "books" of third-hand narrative stories (Gospels) are witnessed accounts, or epistles (letters) were written by the disciples themselves, or anything close to witnessed accounts of the events. The truth that Christianity doesn't want to hear is that only 7 letters of Paul were actually written by the person that the 'books' of the New Test claims to be from, Apostle Paul, the number one heretic in the eyes of the early Jewish Christian groups.

One of the main problems of getting a Christian audience to pay attention to what scholars have discovered is because the vast majority of New Testament scholars "have traditionally seen their role as contributing to a theological enterprise, [what I call the "Christian" or "religious industrial complex"] a clarification of Christian origins that support Christian belief. Mack continues to make the point that the "traditional view of Christian origins has prevailed even in scholarly circles." I am a fan of Bart Ehrman's books but I have been disappointed with him for not writing about certain facts I have uncovered. It's as if he doesn't want to ruffle the feathers of his own theological enterprise more than he already has or he is in denial about the immortal soul doctrine. He could have at least mentioned how no Jewish writer would write "Hades" as a place where dead spirits go. That one doesn't take a college professor to figure out. To common sense logic, it would be the best indication that the Gospels were not written by the disciples. But, then, what do I know? I am not a professor of religious studies.

Furthermore, all of the events portrayed in the narrative gospels of the New Test are not backed up by the Q writings. Although we don't have the full Gospel of Q, other than the snippets found in the Gospels, we don't have any historians from the time-period saying anything about the embellished events from the Gospels. And there are so many flagrant deceptions and outright lies in the narrative gospels that it wouldn't make sense to find an earlier account that includes those same lies.

Mack continues, "Thus the story of Q demonstrates that the narrative gospels have no claim as historical accounts. The gospels are imaginative creations whose textual resources and social occasions can be identified. The reasons for their compositions can be explained. They are documents of intellectual labor normal for people in the process of experimental group formation. Q positions the gospels as period pieces from later phases of the Jesus movements and thus challenges the traditional imaginations of Christian origins as portrayed in the narrative gospels."

I can personally attest to the fictitious nature of the Gospels. After a lifetime of studying to see what is real and what's fake about them, they seem vastly faked for propaganda purposes.

"The narrative gospels can no longer be read as the records of historical events that generated Christianity."

Yes, many of us have known that.

But in spite of that statement, Christianity is still ignoring what agnostic, secular, historian scholars have discovered about the Gospels. By ignoring the implications of what the early Jesus movement really was, an experiment in socialized religion that put emphasis on the well-being of everyone, they can continue to make a comfortable living off of gullible believers.

That whole "Salvation" angle in the NT writings are not even slightly implied in the earliest Jesus movements that produced Q. The story of the Passion and resurrection was the result of "myth-making" by using the Teacher/Founder of Q to create a "layered history of imaginative embellishments of a founder figure, not historical reminiscence, not a meditation on the way in which spiritual life was generated from a crucifixion."

If one reads any of the Gospels with an actively critical thinking-cap on, you can't help seeing not just the fictitiously embellished nature of its composition but the remarkably varied differences between all four of those narrative gospel stories. The only reason for the similarities between the Synoptic Gospels (Matthew, Mark, and Luke) is a combination of Matthew and Luke using parts of Mark's Gospel as a source in their own versions and because of the combined use by all three in the Q sayings. So, when a very different John's 4th Gospel was written, influenced by several different Christologies including some of the previously written Gospels, around the year 100CE, we must reject the claim by Christianity that those accounts of what happened are believable, trustworthy, or anything close to being the plain truth.

With all of the facts presented above, we can't rule out the possibility that Jesus was a real human being, a Teacher of sorts, not a "messiah" ("Christ" in Greek), or someone sacrificed to a god like a Passover lamb, or a "Rabbi" (he wasn't old enough), or a "Holy Ghost" (that was just dumb), or a third branch in a Trinity of three deities (dumber) — but with Mack's research into the "Q" gospel ("Q" comes from the German word for "source") we see evidence for Acts 4:32-35 where the first followers were living a completely socialist commune type of group/community. They were openly sharing the collective wealth of the group with "any who had need."

Where would Jesus get that idea from? Maybe it was his combined experiences with apocalyptic sects like the Essene community.

(*Webster's Ninth New Collegiate Dictionary,* 1986: "members of a monastic brotherhood of Jews in Palestine from 2d century BC to 2d century AD")

Or, as a member of the Pharisees, or his familiarity with the Temple Priests, or even his failed career as a carpenter. At one point, he could have been a follower of John the Baptist and also from having a brother in "James," who was well known throughout Judea for centuries for being "James the Just" or "James the Righteous." And we can understand now that "righteous" and "Just" can either be a strict interpreter of Jewish law or having a "right-wisdom" of someone who understood that a real Designer of the universe would be a purely benevolent deity rather than the tyrannical, authoritarian, bloodthirsty war-lord type of personage that their ancestors had envisioned long before.

In any case, John the Baptist could have put the idea into Jesus' mind that the tree of life could be available to him to be "adopted" by God, which was how the Orthodox writers described it, or to fulfill what Adam failed to do in God's intention to "make man in Our image, after our likeness." Now we can conclude that of the Q sayings that are the most benevolent and altruistic more than likely came from a real Jesus. But even if they didn't come from a real Jesus, I want to assume that we are meant to have them as a guideline to the future of the human race, of which without them our chances of building a peaceful world from the ashes of this dystopian nightmare is nearly futile.

Now, look again at the most blatantly fictitious passages in the New Test with the NKJV, on "Pilate's Decision" in John 18:28-40 and 19:1- 16; showing the extremes that all four narrative gospel writers went to in order to make Pilate look like he was on Jesus' side against the crowd of hateful "Jews." This was the "Jews" who wanted Jesus dead for seemingly no particular reason. If that wasn't obvious enough for you, look at the similarities to that in Luke's version in Luke 23:13-23. Again, making it look like it was the "Jews" who wanted Jesus dead and "turned over to" to do "their will." In Mark we see a shorter version but it says the same thing. Then in the Matthew version, we see the same things but with a twist, Matthew 27:24-25, "When Pilate saw that he could not prevail at all, but rather that a tumult [NIV "uproar" RSV "riot"] was rising, he took water and washed his hands before the multitude [NIV "in front of the crowd"], saying, 'I am innocent of the blood on this just Person [NIV "of this man's blood"]. You see it." And all the people answered and said, "His blood be on us and on our children." A translation called "New English Bible" has, "And with one voice the people cried, "His blood be on us, and on our children." The "Living Bible" has the crowd as a "mob." And a translation called the "Phillips Modern English" has "the whole crowd replied…"

The truth is more believable. As stated above, Pontius Pilate had so many men executed without a proper Roman-based trial (or any reasonable trial for that matter) that a future Emperor of historical significance, Caligula, had Pilate arrested, brought back to Rome, and sentenced to commit suicide for his crimes. Pilate did not try to release Jesus at all, that is all lies to make the Jewish people fighting a war against Roman occupation look as evil as they could imagine. From what we know historically about Pilate, it would be too much of a stretch of one's imagination to consider the narrative gospel accounts here as true. That alone would have been unheard of. The Greek Christians of Pauline Christology who wrote the narrative gospels were doing so during the Jewish war for independence against the vast Roman military. These gospels did not originally have a title or author. Maybe it was a custom of writers to leave their work anonymous, but I very much doubt that as the reason. It was to protect the writers from the wrath of the Roman army. Even though they all went out of their way to fabricate the blame for the crucifixion on "the Jews," that was no guarantee that those Pauline Christian, Greco-Roman writers were safe from Roman reprisal for writing about a crucified Jewish messianic figure by the Roman authority.

If the Jesus of real life was benevolent in his desire for a socialist "Way" of life, and Marcion was sure that Jesus' father was not the god of the Jews, we can safely assume that the parts of the narrative gospels which have Jesus saying altruistic things is reasonable. Judaism changed after the Jewish war. They, as all religions, adopted the immortal soul doctrine and assumed that humans are God's children — literally.

Paul/Saul and the narrative gospels are directly to blame for two disastrous legacies. Their licentious "Salvation through faith" that eliminated the keeping of (virtually) anyone's laws including the commandments and statutes {for a free ride to heaven} and the genocidal, antisemitic history behind Christianity. How does Christianity deal with all this? Business as usual, continued deception, bigotry, greed, and hateful rhetoric.

So, what was Apostle Paul's motivation in all this? Was it to prevent the Jewish war against Rome. No, he only went after the Jesus followers, even to the point of infiltrating their group. And no matter what Acts says about it, Saul was more involved in the stoning of Stephen than just holding the "cloaks," "clothes," or "garments" of Stephen's murderers and then "consenting" or "approving" of the murder. Some honest scholars will tell you that Paul/Saul was a lunatic who thought that Jesus was originally an angel sent from heaven to fulfill the purpose of dying for mankind's sins as a human sacrifice to a god. You know, that nonsense.

Christianity is in too deep to eliminate the New Testament as their infallible, inspired word of God. So they won't. And their desire to see the "end of the world" ("age") could spell out the epitaph for the human race. They are pushing world events in that direction instead of making peace a priority.

So, back to the question of Apostle Paul's main motivation for his legacy. It wasn't the voice that spoke to him on the road to Damascus, that was nonsense, obviously a lie to infiltrate the remnants of the Jesus movement in Jerusalem. If we could ask my comfortably wealthy brother-in-law minister what Saul/Paul's main motivation was, he might say, "Tithes and offerings to line his pockets with." But he won't say anything like that.

Did Saul/Paul believe his own dogmatic rhetoric of a "second coming" of Jesus to "meet the Lord in the air" as the kingdom of God comes down from the "clouds of heaven"? If he did, he was even more delusional than I thought and will be considered the main architect of the impending Sixth Mass Extinction Event. Why? As the majority of all religions await a messiah of some kind, they're not doing anything about the precarious dystopian nightmare we all find ourselves in. It's too late now to stop climate change. We are either approaching or at 9 billion people and resources everywhere are being stretched to beyond their limits. As religion needs to keep people ignorant to fuel their business, the further evolving of human physiology and consciousness on this planet is in dire jeopardy.

If what you did with your life leaves a legacy behind of being mindful of the needs of others, you followed in the footprints of a real, once alive, completely human Jesus of Nazareth. That is the number one manifestation of a one true Designer of the universe, an Impersonal Universal Consciousness. But IT is not a god or God of anyone's religious beliefs. It's a deity who's waiting for us to either evolve into a benevolent nature from our old survival-of-the-richest animalistic nature or we destroy our species from existence. It's such a large universe that there are surely other planets where life has evolved into our species and facing the same threat of extinction but might overcome that destructive nature and successfully evolve into the highest level designed to achieve.

59

The Ebionites

Uncovering hidden proofs of Christianity's Jewish roots reveals that the coming "Christ" (if it happens) will be a thoroughly "Jewish" and the originally human Jesus, not according to the Bible or religion but factually accurate Jewish Christian history and textual criticism of the New Testament. However, even if Jesus is not the "firstborn" in this hypothesis and the entire Bible and all religion is just a whole lot of human imagination, the hypothesis of an evolution process within this designed universe for some kind of higher life-form still makes sense. We're still evolving. What would be a more logical purpose for the design of the universe than evolving this material intelligent life into a higher advanced form through an (eventual) transmutation? Why wouldn't this all be a design to emerge and advance life to its highest possible potential? This really does make more sense than anything else. Why don't more people, anyone other than me, see this reality?

"Lost Christianities: Books that Did Not Make It into the New Testament," by Bart D. Ehrman, pg. 100-101, "Proto-orthodox authors clearly agree that the Ebionites were and understood themselves to be Jewish followers of Jesus. They were not the only group of Jewish-Christians known to have existed at the time, but they were the group that generated some of the greatest opposition. The Ebionite Christians that we are best informed about believed that Jesus was the Jewish Messiah sent from the Jewish God to the Jewish people in fulfillment of the Jewish Scriptures.

They also believed that to belong to the people of God, one needed to be Jewish. As a result, they insisted on observing the Sabbath, keeping kosher, and circumcising all males. That sounds very much like the position taken by the opponents of Paul in Galatia. It may be that the Ebionite Christians were their descendants, physical or spiritual. An early source, Irenaeus, also reports that the Ebionites continued to reverence Jerusalem, evidently by praying in its direction during their daily acts of worship."

"The Ebionites' Jewishness did not endear them to most other Christians, who believed that Jesus allowed them to bypass the requirements of the Law for salvation. The Ebionites, however, maintained that their views were authorized by the original disciples, especially by Peter and Jesus' own brother, James, the head of the Jerusalem church after the resurrection.

One other aspect of the Ebionites' Christianity that set it apart from that of most other Christian groups was their understanding of who Jesus was. The Ebionites did not subscribe to the notion of Jesus' preexistence or his virgin birth. These ideas were originally distinct from each other. The two New Testament Gospels that speak of Jesus being conceived of a virgin (Matthew and Luke) do not indicate that he existed prior to his birth, just as the New Testament books that appear to presuppose his preexistence (cf. John 1:1-3, 18; Phil.2:5-11) never mention his virgin birth. But when all these books came to be included in the New Testament, both notions came to be affirmed simultaneously, so that Jesus was widely thought of as having been with God in eternity past (John; Paul) who became flesh (John) by being born of the Virgin Mary (Matthew and Luke).

Ebionite Christians, however, did not have our New Testament and understood Jesus differently. For them, Jesus was the Son of God not because of his divine nature or virgin birth but because of his "adoption" by God to be his son. This kind of Christology is, accordingly, sometimes called "adoptionist." To express the matter more fully, the Ebionites believed that Jesus was a real flesh-and-blood human like the rest of us, born as the eldest son of the sexual union of his parents, Joseph and Mary. What set Jesus apart from all other people was that he kept God's law perfectly and so was the most righteous man on earth. As such, God chose him to be his son and assigned to him a special mission.

It appears that Ebionite Christians also believed that since Jesus was the perfect, ultimate, final sacrifice for sins, there was no longer any need for the ritual sacrifice of animals. Jewish sacrifices, therefore, were understood to be a temporary and imperfect measure provided by God to atone for sins until the perfect atoning sacrifice could be made. As a result, if these (Christian) Jews were in existence before the destruction of the Jewish Temple in 70 CE, they would not have participated in its cultic Jewish practices; later they, or at least some of them, evidently remained vegetarian, since in the ancient world the slaughter of animals for meat was almost always done in the context of a cultic act of worship."

Amazing right? Unfortunately for us, there are no complete texts of writings by early Jewish-Christians that survived the Orthodox institution's rise to power. The "Nag Hammadi Library" was sets of ancient Christian documents discovered in Egypt in 1945. Predominantly Gnostic in nature, they attest to a time in Christian history when any writings which could be considered contrary to orthodox doctrines was destroyed or, in its case, hidden away for safe keeping. The word "orthodoxy" relates "right doctrine", while the word "heresy" related "wrong doctrine." A "heretic" was considered anyone who taught anything that was not compatible with the established views that the Romans would allow. Of course, the Jewish Christian teachings were so much contrary to the dominant Roman and Greek "Pauline" teachings of the Christian world that even though many other "rejected" writings did survive of other Christologies of this period in history, the writings of Jewish Christians, the Ebionites and the "Nazareans," are only available in quotes by early church "fathers."

T*HE GNOSTIC SOCIETY LIBRARY,* about "The Nag Hammadi Library, a collection of thirteen ancient books (called "codices") containing over fifty texts, was discovered in upper Egypt in 1945. This immensely important discovery includes a large number of primary "Gnostic Gospels" – texts once thought to have been entirely destroyed during the early

Christian struggle to define "orthodoxy" – scriptures such as the Gospel of Thomas, the Gospel of Philip, and the Gospel of Truth. The discovery and translation of the Nag Hammadi library, initially completed in the 1970's, has provided impetus to a major re-evaluation of early Christian history and the nature of Gnosticism."

It was "heresy-hunters" of Orthodox Christianity who believed that it was their duty to stamp out, to burn, or at least discredit anything written by other early Christologies or especially any Jewish Christology. Among those there were not any Jewish Christological writings. However, Bart knows where some Jewish Christological writings remains and those passages are in pieces that were written about by early "church fathers" of the Orthodox Church instituted by Roman Emperor Constantine and beyond.

LOST SCRIPTURES: Books that Did Not Make It into the New Testament

> *Bart D. Ehrman* Pg 15), "The Gospel according to the Hebrews is quoted by a number of church fathers... written... probably during the first half of the second century. Regrettably, the book no longer survives intact, but only in the scattered references to it in these other authors' writings. Its name probably derives from the circumstance that it was used principally by Jewish-Christians... it was called this by outsiders... not by those who actually used it. The Gospel according to the Hebrews was written in Greek and narrated important events of Jesus' life, including his baptism,

> temptation, and resurrection. It appears, however, that these stories were not simply taken over and modified from the Gospels that came to be included in the New Testament. They were instead alternative forms of these traditions that had been passed along orally until the unknown author of this Gospel heard them and wrote them down."

I'll repeat some lines from that to let it soak in by asking you this question… If you wanted to know how we can figure out what a real historical Jesus might likely have said that didn't get into the New Testament… after what I have revealed here…

> "… it was used principally by Jewish-Christians. It appears, however, that these stories were not simply taken over and modified from the Gospels that came to be included in the New Testament. They were instead alternative forms of these traditions that had been passed along orally until the unknown author of this Gospel heard them and wrote them down." "these stories were not simply taken over and modified from the Gospels that came to be included in the New Testament."

Even though Gnostic Christian beliefs were different in some ways from other Christologies, the Jewish Christians and Gnostic Christians had both separated from the Pauline Christology but are closer together in some ways where you might find a small village during 2nd to early 4th centuries the term "Gnostic" itself means knowledge of something unknown, or just "secret-teachings" or "hidden-knowledge," which is what the Way teachings of Jesus was becoming and eventually faded into obscurity, becoming hidden until now. Literally a "hidden knowledge."

And as we have seen, the purpose and plan of the Intelligent Designer has been very much a hidden knowledge to us. There are seven quotations of this Jewish-Christian "Gospel" that "survives in our ancient sources." We will examine a few of them to see if there are any parallels to what we've been learning if the real Jesus had this insight into the Intelligent Designer's plan.

1, "As it is also written in the Gospel according to the Hebrews, "The one who is amazed will rule, and the one who rules will find rest." Does that mean that this person will be one who "sleeps" while awaiting a PREDESTINED resurrection? Maybe death will be just a "rest" for those were amazed at the hidden knowledge? That going deeper into speculation.

2, "If anyone accepts the Gospel according to the Hebrews, there the Savior himself says, "Just now my mother, the Holy Spirit, took me..."

If this Gospel claims that the "Holy Spirit" is Jesus' "mother," and we know "God" Himself is referred to many times as his "Father," it might make sense. Because it is the Holy Spirit, the "tree of life," which will impregnate a human being to grow the seed in the same "Way" that Jesus got baptized, then gives birth to a child of God upon the actual "resurrection."

That same analogy could include our own process someday to possibly becoming the children of the Designer of the universe at the end of time. The tree of life would symbolically be the "mother." This is starkly true of our existence. The Universe is what created us — it is how we emerged and evolved. The "Father" would be the Designer Itself. However, and let me make this point clear, our mother and father would need to raise their offspring together with the offspring. So, people who think that they have that seed implanted in them because they were baptized the same way Jesus did, I hate to disappoint you, but Jesus knew all this in as much of his ancient mind come understand. A safeguard in which I am sure the Designer would have in place, designed into the functions and processes, is not giving the seed from the tree of life to anyone who doesn't understand the very ultimate and deepest hidden meaning of our designed evolution from the universe.

So, I doubt that anyone in this world so far since Jesus (maybe) would be carrying that seed. Mother-universe doesn't have the tree of life available to us, which could be called the "Holy Spirit," but doesn't exist to us. God, the Impersonal Universal Conscience, does not exist in our reality. IT does not intervene, fine-tune, guide or direct anything so far as I can tell and possibly can't because it only exists outside of our reality. Outside of our 3-dimensional space/time continuum. And when the holographic program that the universe is runs out of time and energy, I can only assume that a Designer of the Universe will begin a kind of breeding period within a breeding universe with those who are eventually chosen from our space/time continuum. It makes sense to me.

LOST SCRIPTURES by Bart D. Ehrman: The Gospel of the Hebrews, (5) {after the resurrection} "But when the Lord had given the linen cloth to the servant of the priest, he went and appeared to James. For James had taken a vow not to eat bread from the time he drank the cup of the Lord until he should see him raised from among those who sleep." And soon after this it says, "The Lord said, 'Bring a table and bread.'" And immediately it continues, "He took the bread and blessed it, broke it, gave it to James the Just, and said to him, 'My brother, eat your bread. For the Son of Man is risen from among those who sleep.'"

Unfortunately, that part is a mirror image of what the other writings would say. However, makes it appear as if James (the brother of Jesus, the Just) was at the Last Supper and refused to eat bread for those three days that Jesus was dead. And then "soon after this" appears to be when the risen Jesus comes to see his brother. It also points to the fact that they never believed in the immortal soul doctrine, instead, matching what it says in the writings of both Old and New Tests that death is a sleep without dreams awaiting a resurrection (which would be eventually scientifically discovered).

Not going to a heaven or a hell, not at all, that is the result of the mythology influenced Christology based on the Pauline Christianity that began ruling the Roman Empire which later went on to Conquer a good portion of the known world, even as far away as northern sections of Europe, Asia, and South America as time went by. How can people now say that their religious denomination, group, or cult is the right one? They don't know what they don't know at all. So much confusion over whose opinion is right, to them, without knowing the whole truth. The truth of the reality of our existence.

Again, we see the analogy of describing death as a "sleep."

> "*Lost Scriptures*," by Bart Ehrman, "Later in that Gospel that we have mentioned we find the following written, {the baptism of Jesus} "It came to pass that when the Lord came up from the water, the entire fountain of the Holy Spirit descended and rested on him; and it said to him, 'My Son, in all the prophets I have been expecting you to come, that I might rest on you. For you are my rest, you are my firstborn Son, who rules forever."

It was at that moment that Jewish Christians believed that the fully human Jesus was given the fruit of the tree of life. What they thought that meant is a mystery. But we can speculate. Like a seed of the Designer that would grow in him until his physical death, thus adopting him.

On the "Gospel of the Ebionites" we read in one of Ehrman's other books (I'm not sure which one).

> "The Ebionites differed from non-Christian Jews in asserting that Jesus was the sacrifice for the sins of the world and that all other sacrifices had therefore become meaningless. Among other things, this belief led them to embrace a vegetarian diet, since most meat was procured, in the ancient world, through the religious act of sacrificing an animal. One of the sacred books these Jewish Christians appealed to in support of

> their views was known in antiquity as the Gospel of the Ebionites. Regrettably, the book as whole has been lost; but we are fortunate to have some quotations of it in the writings of an opponent of the Ebionites, the fourth-century heresy-hunter, Epiphanius of Salamis... Some of the Ebionites' distinctive concerns are embodied in their Gospel. This is shown, for example, in the reference to the diet of John the Baptist, in which the canonical statement that he ate locusts (i.e., meat) and wild honey was modified by the change of simply one letter, so that now the Baptist, in anticipation of the Ebionites themselves, maintains a vegetarian cuisine: here he is said to have eaten pancakes and wild honey."

Another lost gospel by Jewish Christians is called...

(*Wikipedia*) "The Gospel of the Ebionites is the conventional name given by scholars to an apocryphal gospel extant only as seven brief quotations in a heresiology known as the Panarion, by Epiphanius of Salamis; he misidentified it as the "Hebrew" gospel, believing it to be a truncated and modified version of the Gospel of Matthew. The quotations were embedded in a polemic to point out inconsistencies in the beliefs and practices of a Jewish Christian sect known as the Ebionites relative to Nicene orthodoxy.

The surviving fragments derive from a gospel harmony of the Synoptic Gospels, composed in Greek with various expansions and abridgments reflecting the theology of the writer. Distinctive features include the absence of the virgin birth and of the genealogy of Jesus;

an Adoptionist Christology, in which Jesus is chosen to be God's Son at the time of his Baptism; the abolition of the Jewish sacrifices by Jesus; and an advocacy of vegetarianism. The omission of the genealogical records and the virgin birth of Jesus narrative is explained by Epiphanius as being because 'they insist that Jesus was really man.'

It is believed to have been composed some time during the middle of the 2nd century in or around the region east of the Jordan River. Although the gospel was said to be used by "Ebionites" during the time of the early church, the identity of the group or groups that used it remains a matter of conjecture.

The Gospel of the Ebionites is one of several Jewish–Christian gospels, along with the Gospel of the Hebrews and the Gospel of the Nazarenes; all survive only as fragments in quotations of the early Church Fathers. Due to their fragmentary state, the relationships, if any, between the Jewish–Christian gospels and a hypothetical original Hebrew Gospel are uncertain and have been a subject of intensive scholarly investigation. The Ebionite gospel has been recognized as distinct from the others, and it has been identified more closely with the lost Gospel of the Twelve. It shows no dependence on the Gospel of John and is similar in nature to the harmonized gospel sayings based on the Synoptic Gospels used by Justin Martyr, although a relationship between them, if any, is uncertain. There is a similarity between the gospel and a source document contained within the Clementine Recognitions (1.27–71), conventionally referred to by scholars as the Ascents of James, with respect to the command to abolish the Jewish sacrifices." (*Wikipedia*)

It can be assumed that by the second century the Jewish Christians were divided into different forms of mutations from some of their original beliefs but in the Jewish tradition of religions you always kept close as you could to the original religious beliefs of your religion no matter how far outside of the traditional beliefs you had ventured. That was not the same for any other Christology as there appeared maybe a hundred others, practically each one for every one of those documents in Nam Hammadi.

So far, we have learned that some of the Jewish Christians were considered "Ebionites". But there were others, and their teachings were always in favor of the view that everyone, Jews and Gentiles, will be required to understand and keep God's laws and statutes when the Day of the Lord happens in accordance with how Jesus and James from their messianic apocalyptic Jewish sect would have understood it.

These Jewish-Christians refrained from eating meat because most meat was procured through the religious act of sacrificing the animal to a god. In this abstention from eating dead animal flesh they may have developed a real empathy for animals, as empathy is what we need to evolve into the benevolent life-form that the designed design of the universe is in the process and function of naturally creating. When I take upon myself the different kinds of vegetarianism, I also found myself being at least more respectful toward animals in general. Developing increasing amounts of empathy and concern for all life and its place in our ecosystem and universe is in reality a requirement to fulfill our purpose, if it is our purpose, to become the literal offspring or hybrid of the Designer of the universe. If there is one.

There were plenty of religions that included sacrifices to gods, not just Judaism. However, to get people to stop killing animals I can see the possibility that the universe would not want us killing our animal kingdom for food when we have "the herb of the field." We are all beasts ,we were all made from the same atoms as everything else. WE ARE LIFE. We are the universe. Why kill any living thing at all if we don't have to? It appears that a vegetarian diet is about the most healthy kind if done in moderation.

(*newsinhealth.nih.gov*) "Vegetarians tend to weigh less than meat-eaters, and to have lower cancer rates. "Evidence also suggests that a vegetarian diet is associated with a lower risk of death from certain heart diseases, and that those who follow a vegetarian diet tend to have lower LDL ["bad"] cholesterol levels," says Engel."

If a Designer's plan doesn't unfold until after this second-coming thing it would be logical for the process of "redemption" to start at that time, not before, if it is even a real thing. Therefore, if the Last Supper even happened, which we have reason with logic to doubt, it was only maybe for his "disciples", his "chosen", to be given the "seed" at that time, no one else then or since. The baptism symbolized washing away the sins of the past in order to plant the Holy Spirit into a human being's human spirit. To basically plant the seed of

God Himself into fresh soil from that tree of life. All for the opportunity to grow within the human spirit. As it says in Genesis, "and eat, and live forever."

So, if the baptism annuls the penalty of the original sin, wouldn't committing sin again after baptism require the same death penalty again? This is what Christianity does not understand about their own religion, so they make excuses. The whole thing just sounds illogical. A blood sacrifice is a barbaric concept. Keeping the laws and statutes the Way Jesus taught is a benevolent way to live. They didn't even call themselves "Jewish Christians." They may have just called themselves "the Way."

What was the ancient Greek human sacrifice ritual? Is it possible that it influenced the stories that came to be written, preserved and collected to be included in the "New Testament canon", and thus decreed and deemed as holy writ? Maybe?

(Rutledge)"In Greek mythology, it was said humans were sacrificed to Zeus at the king of the gods' birthplace on top of Mount Lykaion. In one tale, the king Lycaon turns into a wolf after he sacrifices a baby on the altar there. In another tale, a boy is sacrificed with animals, and all the meat is cooked together."

Wikipedia: "Where did human sacrifice come from?

Funeral human sacrifice was widely practiced in the ancient Chinese state of Qin. According to the Records of the Grand Historian by Han dynasty historian Sima Qian, the practice was started by Duke Wu, the tenth ruler of Qin, who had 66 people buried with him in 678 BCE."

Examples of human sacrifice in historical religion included (*thecollector*), "The European cultures of antiquity were also no stranger to the ultimate religious expression of violence. The Greeks, Romans, Celts, and Germanic peoples all practiced ritual human sacrifice to a certain degree.

The ancient Greeks practiced "pharmakos," whereby cripples, criminals, or slaves would be thrown off a cliff to ward off disaster. This process disappeared in later antiquity as the Greeks moved towards a more civilized society. In Roman society, human sacrifice was rare but better documented. Like the Greeks, human sacrifice was seen as a remnant of a more uncivilized age, and as Rome progressed, the victim was replaced with symbolic items such as dolls or statuettes."

How absurd could it be that a God gave all human beings a set of laws and if we break one we must be put to death? That was what "stoning" a person to death was for. THEY, meaning the religious authority, or just about anyone — were the judge, jury, and executioner of anyone breaking those "Commandments." In order to pay the price for the "wages of sin" one had to die. Which is as barbaric and backwards a belief as any if we would all look at it for what it is.

Romans 6:23, (NRSV) "For the wages of sin is death, but the free gift of God is eternal life in Christ Jesus our Lord."

(*faithward*) "What is the significance of the sacrificial death of Jesus? In the sacrifice of Jesus's crucifixion we are shown the depths of God's love for us and the lengths taken to save us from our sins. And in Jesus's resurrection we see God's triumph over death, pointing toward the promise of eternal life in God's presence."

It just doesn't make sense in our modern world. That is definitely not the Intelligent Designer of the Universe behind that crap.

Some early Christians who were called "Gnostic" believed that Jesus was not even made of matter, just a virtual "image" of a man. Of course, that's almost as ridiculous as the human sacrifice of Jesus for the Salvation by faith to those who believe. That seems like a con game. It's like God saying, "Worship Me or die!" Beware of those who think they are permitted by God to make the rules for everybody. That's exactly what is the most dangerous thing about religion. History is loaded with examples of it. It once again threatens the liberty, democracy and freedom in the world from Islam, Christianity, and Zionist Judaism.

It would have been easy for humans influenced by mythology to accept a half-man/half-god person. However, if the nature of the real Designer is not compatible with material things in a material universe, the "seed" itself of this Designer of the universe is not "divine" at all. What is flesh is flesh, and what is Spirit is Spirit.

JOHN 3:6-8, (NRSV) "What is born of the flesh is flesh, and what is born of the Spirit is spirit. Do not be astonished that I said to you, "You must be born from above." The wind blows where it chooses, and you hear the sound of it, but you do not know where it comes from or where it goes."

So Jesus would have been fully human (flesh) the entire course of his short life (about 30 years old) until after his death (Spirit). The seed became the sprout after he died. That hasn't happened yet, if it ever will. The evidence points to Jesus NOT coming back to life and was probably left to rot on the cross and as a warning to others. The books in the New Testament are filled with lies. Especially the content at the end of the Gospels and all of the book called ACTS.

It's much more logical to view Christianity from the beliefs of the first Jewish Christians who claimed to be the very descendants of the disciples. The problem there is that we only have pieces of writings from early Jewish Christian groups according to ancient Orthodox scholars. It has long been lost to the ages. In that case piecing together what we can assume that they believed appears to be similar to Orthodox in some ways and far different in others. Particularly when it comes to converting to a Christian form of Judaism where Commandments, laws, and statutes included what you can and cannot eat being important but not nearly as important as developing the benevolence that Jesus taught by example and parable.

60

Thought Experiment

Answer these questions (honestly):

Do you believe in a God of religion?

If Yes, the odds are that it is one of the four major religions.

Is it Jewish, Christian, Muslim, or Buddhist?

There are different versions, denominations, and sects. Which one do you believe is the true religion?

Why? Is it A. Because you were raised to believe it?

Or is it B. You don't know why you believe it?

Or C. Someone convinced you to accept it as real by having blind faith?

Whether it is A, B, or C, it is not logical and you are not being honest with yourself. The truth is not what many people admit to. You really don't know if there is a God of any religion or not because no one knows that. You can only assume what you believe if you believe any version of any religion. The odds may be more favorable toward there being an intelligence behind the existence of this universe but no religion represents IT.

The core problems on our planet could easily be solved if religion is eliminated as long as the priority of everyone is to help everyone to survive and thrive. It is called "Humanism." Join the club. Albert Einstein was a member.

On page 136 of one of Bart's books we see the best take on Revelation that I have seen:

> "Armageddon: What the Bible Says About the End of the World" by Bart D. Ehrman "It should be clear by now that the beast of Revelation 13 is not a future ruler of a global government ("the European Union"), an Antichrist yet to come. John is writing for his own day. The beast is first-century Rome, its empire, and its emperor all rolled into one: haughty against God, controller and exploiter of the entire world, worshipped by all nations, and persecutor of Christians. So nothing in the book of revelation before the 19th chapter was written to be prophecy for the future. Even then, all I can get from the last chapters is predominantly nonsensible, as a judgment occurs with horrendous results. Very, very few people, hardly any in the grand scope of things, will end up participating in the "first resurrection. (Rev 20:5-6)""

"Jonathon Kirsch, *A History of the End of the World*:

> "But the contempt for imperial Rome that suffuses the book of Revelation was rendered suddenly and wholly

obsolete when Emperor Constantine (ca. 280-337) embraced Christianity in the early fourth century. Under Constantine and his sons, the Christian Church was elevated from a marginalized and criminalized sect into the favored and protected faith of the imperial family and, eventually, a kind of shadow government whose reach extended throughout the Roman Empire. Once the Roman emperor was a Christian rather than a persecutor of Christians, the condemnation of imperial Rome in the book Revelation no longer made much sense. Indeed, the Christian church now styled itself as "the Church Militant and Triumphant."

"Jonathon Kirsch, "*A History of the End of the World,*"

pg. 27 "Thus, according to the core theology of the Bible, God is the sole author of history and the sole arbiter of what happens to humankind. And so, if God is sufficiently provoked by the stubbornness of the Chosen People to make them suffer famine or pestilence, conquest or exile, it means only that they are getting what they bargained for and what they deserve."

"Science" can be defined as a "systematized knowledge derived from observation and study."(Wikipedia) "Reason" can be defined as, "the ability to think, draw conclusions, etc." and "to think logically; analyze" and "The power of the mind to think, understand, and form judgments by a process of logic (Google)." Britannica says, "think, cogitate, reflect, speculate, deliberate mean to use one's powers of conception, judgment, of inference."

> *MARK 10:17-21*, (NRSV) "Good teacher, what must I do to inherit eternal life?" And Jesus said to him, "Why do you call me good? No one is good but God alone. You know the commandments: You shall not murder; You shall not commit adultery; You shall not steal; You shall not bear false witness; You shall not defraud; Honor your father and mother." He said to him, "Teacher, I have kept all these since my youth." Jesus, looking at him, loved him and said, "You lack one thing; go, sell what you own, and give the money to the poor, and you will have treasure in heaven; then come, follow me."

"Fundamental Beliefs of the United Church of God" by United Church of God, an International Association:

"We believe that Christians are forbidden by the commandment of God from taking human life directly or indirectly and that bearing arms is contrary to this fundamental belief. Therefore, we believe that Christians should not voluntarily become engaged in military service, we believe they shall refuse conscientiously to bear arms."

Perfect. That really impressed me when I read it. I am pretty sure that's what the historical Jesus wanted. Like the one who said, "Blessed are the peacemakers." Forcing a peaceful solution to humanity's problems at the end of a gun or bomb has never and will never produce a lasting peace. I have studied the doctrines of this organization and even though they hold some of the purest doctrines based somewhat on what the original Jewish Christians believed, they also make the mistake of calling the Bible infallible and study Apostle Paul's licentious beliefs as if he represented the truth. To me, it seems very contradictory to hold some of the Jewish Christian traditions while thinking that Paul's doctrine of eliminating everything Jewish about the Jesus movement is inspired by something divine. His alleged conversion event is so obviously faked for the purpose of infiltrating the Nazarene sect of Judaism, the original Jesus sect, to turn them from being apocalyptic and messianic to something that reflects a Roman and Greek mythology that was for the purpose of protecting his beloved Roman Empire from the Jewish war for independence that happened anyway in 66 CE. He fooled a lot of people at that time and ever since.

61

And In The End

These five points add up to the reason for our very existence in this reality.

1. This universe was designed biofriendly to produce life wherever conditions happen to make that possible as it did here. That is the science of astrobiology.

2. Life evolved from natural selection's survival of the fittest through the instinct of self-preservation.

3. Empathy is not compatible with natural selection and interferes with the process of the evolution of species. Empathy is the opposite of and diametrically opposed to the evolutionary process of survival of the fittest/richest.

4. Empathy is a human condition of emotion that did not evolve by natural selection but has been evolving in us after we became consciously self-aware of ourselves and our existence.

5. The universe is designed to produce life and evolve that life from simple forms of single-celled organisms through DNA improvements in complexity to the multi-cellular forms. That

is the universe's primary function, creating living things, wherever conditions are right. In our case, it produced sentient, sapient, and consciously self-aware life that became influenced over time by the emotion of feeling empathy toward other life-forms as the secondary purpose of the universe; evolving us into benevolent life-forms through increasing levels of empathy.

Our existence was designed to produce physical life and evolve that life into the image and likeness of it's Designer.

Empathy is not natural in evolution. Carnivores don't feel empathy for the animals they kill to eat. That would not work in the grand design of a biofriendly universe. Vegetarians do not feel empathy toward the plant life they consume. Life feeds off life. It is the natural way of nature – the self-preservation instinct to consume to give us the energy to continue living is essential to life itself.

So where does empathy itself in human beings come from? It can be an emotional experience to watch a living thing suffer. It appears to be something that is evolving inside of us. Evolution is based on life preserving itself for the sake of itself, not for the sake of other life. Empathy would not be part of an evolving life function unless it is the purpose of the design-- and has been all along – to develop it into our consciously aware human "spirit" or whatever that thing in us is that we feel emotion with or that enables us, but not animals, to know that we are naked. It became part of our very nature to feel empathy (or at least the capacity for it) at some point in our evolution form animals.

Christians have been taught to disregard evolution because of the fantasy of biblical creationism. However, if the evolution process was (is) the one and only, number one, single, solitary purpose of our entire existence and design of the universe, then it has been our religions that are the true enemy of the purpose behind the design and empathy is what we are supposed to be evolving into – a benevolent nature that is the only real representation of a deity from beyond our dimensional plane of existence. Love, people, it's more than just a sentimental feeling or emotional response in us, it is our destiny. If those people who can't feel empathy cannot evolve to the next stage of our journey after death (upon a resurrection), they cannot be of any use to the impersonal universal consciousness and will not be resurrected in the same manner that an altruistic, empathetic individual will be. It would serve no justifiable purpose to resurrect the sociopath just to have them incinerated. Some kind of second chance period would be the only hope for many people who have no empathy.

The official last song sung by the Beatles will have more truth in it than anyone has realized. "And in the end, the love you take is equal to the love you make." Imagine that.

In the final analysis; how could something as diametrically opposed to natural selection as empathy is, come from the same DNA's biological system that produced natural selection, unless empathy was designed to emerge and evolve in the human consciousness?

"*On the Origin of Time: Stephen Hawking's Final Theory*" by Thomas Hertog (page 259):

> "We learn from quantum cosmology that biological evolution and cosmological evolution aren't fundamentally separate phenomena but two vastly different levels of one giant evolutionary tree."

My singular purpose in life has been to discover and write about the enigmatic mystery of why the universe and I exist and present what I find to the world to see. Maybe to understand it and why we are all part of a designed design (rather than a random chance occurrence) in the evolution of this fleeting temporary life of limitations to eventually evolve humankind (and maybe ourselves) to transform into a benevolent and altruistic life-form of pure consciousness who will be able to exist outside of our physical realm into higher dimensions. This book is ready for the world, along with the scrutiny that it will surely receive simply because conventional religion and certain corners of the science world are characterized by closed minds to these questions of existence.

Some of them will gradually learn and adapt, like our primate ancestors did at every key stage of evolution. (and some will not)

IMAGINE

John Lennon

"Imagine there's no heaven
It's easy if you try
No hell below us
Above us, only sky

Imagine all the people
Livin' for today
Ah

Imagine there's no countries
It isn't hard to do
Nothing to kill or die for
And no religion too [to kill or die for]

Imagine all the people
Livin' life in peace
You

You may say I'm a dreamer
But I'm not the only one
I hope someday you'll join us
And the world will be as one

Imagine no possessions
I wonder if you can
No need for greed or hunger
A brotherhood of man

Imagine all the people
Sharing all the world
You

You may say I'm a dreamer

But I'm not the only one
I hope someday you'll join us
And the world will live as one"

Bibliography

"Alive and Well with Wild Food" by Peter Wagner (2006 "Example Product Manufacture")

"Alone in the Universe: Why Are Planet is Unique" by John Gribbin (2011 "Trade Paper Press")

"A History of the End of the World" by Jonathan Kirsch (2006 "Harper One")

"American Standard Version (ASV) (1901 "Lockman Foundation")

"A Natural History of the Future" by Rob Dunn (2021 "Hatchet Books Group")

"Armageddon: What the Bible Says About the End of the World" by Bart D. Ehrman (2023)

"Astrobiology: Exploring Life in the Universe" by Lewis Dartnell (2011 "Rosen Publishing Group"))

"Britannica"

"Darwinism Under the Microscope" by Gills and Woodward (2002 "Charismas House")

"Did Jesus Exist: The Historical Argument for Jesus Christ" Bart D. Ehrman (2012 "Harper Collins")

"Encyclopedia of Word and Phrase Origins" by Robert Hendrickson (QPB, Facts On File, 2004)

"Excavating Jesus: Beneath the Stones, Behind the Texts" by Crossan and Reed (2001 "Harper One")

"Four Views on Creation, Evolution and Intelligent Design" (2017 by "Zondervan Press")

"Eight Translation New Testament" (1977 "Tyndale")

"God is Not Great" Christopher Hitchens (2007 "Hatchet Group")

"God's Problem" by Bart D. Ehrman (2008 "Harper Collins")

"Heading Toward Omega" by Kenneth Ring (1984 "William Morrow & Company")

"How Jesus Became God: The Exaltation of a Jewish Preacher from Galilee" by Bart D. Ehrman (2014)

"James the Just and Christian Origins" by Craig Evans and Bruce Chilton (1999 "Chilton")

"Jesus Christ the Real Story" by United Church of God (2012)

"Lost Christianities and the Faiths We Never Knew" by Bart D. Ehrman (2002 "Oxford")

"Lost Scriptures: Books That Did Not Make It Into the New Testament" by Bart D. Ehrman (2003)

"Merriam/Webster's Deluxe Dictionary"

"Misquoting Jesus: The Story of Who Changed the Bible and Why" by Bart D. Ehrman (2005 "Harper")

"New Roget's Thesaurus" (1991)

"On the Origin of Time: Stephan Hawking's Final Theory" (2023) by Thomas Hertog

"Rare Earth: Why Complex Life is Uncommon in the Universe" by Ward & Brownlee (2000 Capernicus)

"Tanach" {Jewish Bible} The ArtScroll Series (1997)

"The Blind Watchmaker" by Richard Dawkins (1986 Longman Scientific Technical)

"THE BROTHER OF JESUS" by Hershel Shanks & Ben Witherton lll (2003 Harper One)

"The Brother of Jesus: and the Lost Teachings of Christianity" by Jeffrey J Bütz (2005 Inner Traditions)

"The Case for a Creator" by Lee Strobel (2004 Zondervan)

"The Edge of Evolution: The Search for the Limits of Darwinism" by Michael J Behe (2007 Simon/Shuster)

"The Gnostic Society Library" (1995)

"The Good News Bible" (GNB, 1976)

"The Holographic Universe" by Michael Talbot (1991 Harper Perennial)

"The Holy Bible: King James Version" (KJV, 1611)

"The Holy Bible: New International Version" (NIV) {2011}

"The Holy Bible: New King James Version" (NKJV)

"The Holy Bible: New Revised Standard Version" (NRSV)

"The Holy Bible: New Revised Standard Version Updated Edition" (NRSVUE)

"The Holy Bible: Revised Standard Version" (RSV)

"The Human Question: What People Believe About Evolution" by (2003) Hervey Cunningham Peoples

"The Jerusalem Bible: The New Testament" (Double Day 1968)

"The Journey of Man: A Genetic Odyssey" by Spencer Wells (2003 Princeton University Press)

"The Lost Gospel: The Book of Q and Christian Origins" by Burton L. Mack (1993)

"The Misunderstood Jew: The Church and the Scandal of the Jewish Jesus' by Amy Jill Levine (2006)

"The Mythmaker: Paul and the Invention of Christianity", by Hyam Maccoby (1986)

"The New American Webster's Handy College Dictionary: Fourth Edition" (1956 Berkeley Books)

"The New Testament" by Bart D. Ehrman (1996 Oxford University Press)

"The Oxford Annotated Bible: Revised Standard Version College Edition"(1962)

"The Politically Incorrect Guide to Darwinism and Intelligent Design" by Jonathon Wells PhD (2006)

"The Religious Experience" by Ninian Smart (1969 Charles Scribner & Sons)

"The Sixth Mass Extinction: An Unnatural History" by Elizabeth Kolbert (2014 Henry Holt)

"Webster's Handy College Dictionary Fourth Edition

"Webster's Expanded Dictionary 1995 Edition"

"Webster's Ninth New Collegiate Dictionary" (1986)

"Webster's New Twentieth Century Dictionary (1966, World Publishing Company)

About the Author

Co-author of (2000, Skyhorse Publishing) "Foraging for Survival: Edible Wild Plants of North America" with Mykel Hawke -at 70 years old- Douglas D. Boudreau has compiled many diverse files on his "Four Essential Subjects of Life: Science, Religion, Survival, and Nutrition." After open-heart surgery 4 years ago, he became obsessed with finding the answers to the mysteries of life and a logical purpose behind the existence of the universe."

The author welcomes thoughtful conversations with informed people. Any reader who wishes to contact him can do so by calling 813-845-4762 and leaving a message with their name, contact number and subject of interest.

www.ingramcontent.com/pod-product-compliance
Lightning Source LLC
LaVergne TN
LVHW040514210926
848480LV00005BA/1450